Pergamon International Library
of Science, Technology, Engineering and Social Studies
The 1000-volume original paperback library in aid of education,
industrial training and the enjoyment of leisure
Publisher: Robert Maxwell, M

A Manual of Chemical and Biological Methods for Seawater Analysis

ERRATA

Sections C, D, E at the bottom of page 141 and the top of page 142 should appear between Sections B and F on page 143.

Parsons   0302874 F   0302882 H

Related Pergamon Titles of Interest

Books

BARNES
Oceanography and Marine Biology* (An Annual Review Series)

BEER
Environmental Oceanography

GORSHKOV
World Ocean Atlas*
Volume 1: Pacific Ocean
Volume 2: Atlantic and Indian Oceans
Volume 3: Arctic Ocean

PARSONS *et al*
Biological Oceanographic Processes, 3rd edition

PICKARD & EMERY
Descriptive Physical Oceanography, 4th edition

POND & PICKARD
Introductory Dynamic Oceanography, 2nd edition

RAYMONT
Plankton and Productivity in the Oceans, 2nd edition (2 Vols)

Journals

Continental Shelf Research

Deep-Sea Research

Progress in Oceanography

Full details of all Pergamon publications/free specimen copy of any Pergamon journal available on request from your nearest Pergamon office

* Not available under the terms of the Pergamon textbook inspection copy service

A Manual of Chemical and Biological Methods for Seawater Analysis

TIMOTHY R. PARSONS

Department of Oceanography, University of British Columbia
Vancouver, B.C., Canada V6T 1W5

YOSHIAKI MAITA

Research Institute of North Pacific Fisheries
Faculty of Fisheries, Hokkaido University, Hakodate, Hokkaido, Japan

and

CAROL M. LALLI

Department of Zoology, University of British Columbia,
Vancouver, B.C., Canada V6T 1W5

PERGAMON PRESS
OXFORD · NEW YORK · TORONTO · SYDNEY · PARIS · FRANKFURT

U.K.	Pergamon Press Ltd., Headington Hill Hall, Oxford OX3 0BW, England
U.S.A.	Pergamon Press Inc., Maxwell, Fairview Park, Elmsford, New York 10523, U.S.A.
CANADA	Pergamon Press Canada Ltd., Suite 104, 150 Consumers Rd., Willowdale, Ontario M2J 1P9, Canada
AUSTRALIA	Pergamon Press (Aust.) Pty Ltd. P.O. Box 544, Potts Point, N.S.W. 2011, Australia
FRANCE	Pergamon Press SARL, 24 rue des Ecoles, 75240 Paris, Cedex 05, France
FEDERAL REPUBLIC OF GERMANY	Pergamon Press GmbH, Hammerweg 6, D-6242 KronbergTaunus, Federal Republic of Germany

First edition 1984

Library of Congress Cataloging in Publication Data

Parsons, Timothy Richard, 1932–
A manual of chemical and biological methods for seawater analysis.
(Pergamon international library of science, technology, engineering, and social studies)
1. Seawater – Analysis. I. Maita, Yoshiaki, 1937– II. Lalli, Carol M. III. Title. IV. Series.
GC101.2.P37   1984   551.46'01   83-23717

British Library Cataloguing in Publication Data

Parsons, Timothy R.
A manual of chemical and biological methods for seawater analysis.
1. Sea-water—Analysis
I. Title   II. Maita, Yoshiaki   III. Lalli, Carol M.
551.46'01   GC116

ISBN 0-08-030288-2 (Hardcover)
ISBN 0-08-030287-4 (Flexicover)

Printed in Great Britain by A. Wheaton & Co. Ltd. Exeter

Preface

The following text is intended to serve as an introduction to the quantitative analysis of seawater. Biological and chemical techniques are described in detail and these are believed to be among those most often used by biological oceanographers. In general, the techniques require a minimum of prior professional training; in addition, methods requiring the use of very expensive equipment have been avoided. As such, it is intended that the techniques will be useful to students, environmentalists and engineers as well as to some other oceanographic disciplines.

The leadership of the late J.D.H. Strickland in publishing *A Practical Handbook of Seawater Analysis* with the senior author of this text is acknowledged, together with the publisher of the former text, The Fisheries Research Board of Canada. The style in which many methods are described here is similar to that in the *Practical Handbook* and some methods are modified from the former book. New methods have been described, particularly with respect to biology; in addition, a section has been written on terms and equivalents which we believe will be useful to many biologists.

Acknowledgments

The authors are grateful to Ms. H. Dovey who prepared most of the figures and to Dr. E.V. Grill who prepared Tables 4, 5 and 6. Dr. Grill also kindly read and commented on the text to the authors prior to its submission for publication. We also wish to thank Mr. H. Yoshida who helped in the testing and modification of several methodologies. The work was sponsored by the National Science and Engineering Research Council in direct response to a request from the awards office.

Caution

The authors strongly recommend that laboratory safety procedures should be followed where necessary as prescribed in such references as *Hazards in the Chemical Laboratory*, *Ed.* L. Bretherick, *Publ.* Royal Soc. Chem. (London), 3rd edition, 1981, 569 pp. Such safety procedures are especially important when using strong acids, alkalis, oxidizing reagents, biological stains, preservatives and radioactive materials.

Contents

List of Figures

List of Tables

General Notes on Analytical Techniques

The following techniques can generally be employed with a minimum outlay of capital equipment. Thus the use of a spectrophotometer, fluorometer, microscopes, Coulter Counter and scintillation counter will cover most of the methods. The use of more sophisticated equipment, such as gas/liquid chromatograms, atomic absorption analyzers and mass spectrometers, is not described since the operation of these pieces of equipment is usually specialized and well described by company brochures. In addition, however, some techniques are not described because there is equipment available which specifically measures the property without requiring further detailed explanations. Such equipment includes salinometers, light meters and Autoanalyzers® (the latter being extensively adapted for nutrient analyses using basic colorimetric techniques which are reproduced here). In other cases, the measurement of a property may still be controversial and also require more expensive equipment; this appears to be the case with dissolved organic carbon (DOC) as discussed, for example, by Gershey *et al.* (*Mar. Chem.*, **7**: 289, 1979).

The techniques described here make considerable use of different kinds of filters – our use of these filters is often made on the basis of the most readily available, such as Millipore® filters; but for some purposes, the polycarbonate Nuclepore® filters or glass fiber filters are superior to the cellulose nitrate filters.

In nearly all the analyses some indication of the statistical reliability of the technique is given such that the confidence limits quoted represent $\pm 2\sigma$, and this is given so that the mean of n determinations will be $\pm 2\sigma \sqrt{n}$ with a confidence of 95%.

In the spectrophotometric methods, the cell: cell blank of cuvettes is not described as part of the procedure because it is assumed that samples, standard and reagent blank will all be measured in a single sample cuvette. The cell: cell blank then becomes part of the reagent blank. However, this procedure should not preclude an independent examination of the variability in reagent blanks by the analyst, particularly in respect to low level detection of certain nutrients.

The text does not entirely follow the International System of Units for the same reason that many scientists do not. Thus, while the basic unit of volume

xiii

in the SI system is m³, it is still preferable from a practical point of view to use the allowed unit of a liter rather than the dm³. The expression of many concentrations in terms of amount per liter further leads to a difference of a few percentage points (usually < 3%) between this concentration (e.g., for a nutrient in μg-at/l) and the salinity which is defined as the salts present in a kilogram of seawater. For convenience of the analyst, however, this kind of difference is small relative to the order of magnitude changes that occur in the biological and chemical properties of seawater.

Methods given in this text are described in terms of procedures and not in terms of interpretation of results. The latter is up to the individual investigator; in some methods, reference material quoted may assist in the interpretation of results.

SECTION 1

Nutrients

1.1. Determination of Nitrate

Introduction

The following procedure is based on a method by Morris and Riley (*Anal. Chim. Acta*, **29**: 272, 1963) with some modifications suggested by Grasshoff (*Kiel. Meeresforsch.*, **20**: 5, 1964) and Wood *et al.* (*J. mar. biol. Assoc.*, U.K., **47**: 23, 1967).

Method

A. Capabilities

 Range: 0.05–45 µg-at/l
 Precision at the 20 µg-at/l level:
  The correct value lies in the range, mean of n determinations $\pm 0.5/n^{\frac{1}{2}}$ µg-at/l (using a 1-cm cell). For nitrate values of < 1 µg-at/l, use of a 10-cm cell is advised.

E. Outline of method

Nitrate in seawater is reduced almost quantitatively to nitrite when a sample is run through a column containing cadmium filings coated with metallic copper. The nitrite produced is determined by diazotizing with sulfanilamide and coupling with N-(1-naphthyl)-ethylenediamine to form a highly colored azo dye which can be measured spectrophotometrically. Any nitrite initially present in the sample must be corrected for.

C. Special apparatus and equipment

 50-ml graduated cylinders
 125-ml Erlenmeyer flasks
 A reduction column may be conveniently made according to Fig. 1 (Note a).

D. Sampling procedure and storage

100 ± 2 ml of seawater should be measured into a 125-ml Erlenmeyer flask and analyzed immediately. If samples are stored, they should be frozen at

3

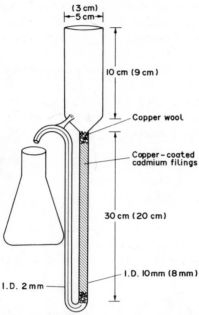

FIGURE 1. Dimensions of a nitrate reduction column. Numbers shown in brackets are for smaller columns which are especially useful for ammonia determination (see Method 2.1).

$-20°C$. In the presence of high concentrations of phytoplankton, samples should be filtered before analysis.

E. Special reagents

1. Concentrated ammonium chloride solution

Dissolve 125 g of analytical reagent grade ammonium chloride in 500 ml of distilled water and store in a glass or plastic bottle.

2. Dilute ammonium chloride solution

Dilute 50 ml of concentrated ammonium chloride solution to 2000 ml with distilled water. Store the solution in a glass or plastic bottle.

3. Cadmium–copper filings

Cadmium filings of a specific size range are required for the columns. They may be bought or made from cadmium metal by filing the metal with a coarse wood file. Filings should pass 2-mm sieve openings and be retained by 0.5-mm openings. Stir about 100 g of filings (sufficient for 2 columns) with 500 ml of 2% w/v solution of copper sulfate ($CuSO_4 \cdot 5H_2O$) until the blue color has left the solution. Place a small plug of copper wool (turnings) in the bottom of the reduction column and fill the column with dilute ammonium chloride solution. Pour in a slurry of the cadmium–copper filings and gently pack the column to a height of about 30 cm. Do not allow the filings to become dried out during the procedure; they should continue to be covered with dilute ammonium chloride, or the seawater samples, at all times. Wash the column thoroughly with dilute ammonium chloride and adjust the flow rate by tapping the side of the column so that about 100 ml is collected in 8 to 12 min. If the flow rate is slower than this, the column has to be repacked. Add a small plug of copper wool to the top of the column.

The cadmium–copper filings may be reactivated after continued use (as judged by the *F* value obtained for the standard). Filings are removed from the column, washed with 5% v/v hydrochloric acid and then washed with distilled water until the pH of the decanted solution is > 5. The filings can then be reactivated with copper sulfate using the procedure given above.

4. Sulfanilamide solution

Dissolve 5 g of sulfanilamide in a mixture of 50 ml of concentrated hydrochloric acid (sp. gr. 1.18) and about 300 ml of distilled water. Dilute to 500 ml with water. The solution is stable for many months.

5. N-(1-naphthyl)-ethylenediamine dihydrochloride solution

Dissolve 0.5 g of the dihydrochloride in 500 ml of distilled water. Store the solution in a dark bottle. The solution should be renewed once a month or directly a strong brown coloration develops.

F. Experimental procedure

1. Add 2.0 ml of concentrated ammonium chloride to the sample in the Erlenmeyer flask. Mix the solution and pour about 5 ml onto the top of the column and allow it to pass through.

2. Add the remainder of the sample to the column and place the drained Erlenmeyer flask under the collection tube (see Fig. 1). Collect about 40 ml and discard; collect 50 ml in a graduated cylinder and dispense this into the Erlenmeyer flask which contained the original sample. Allow the column to drain before adding the next 5 ml sample (1 above).

3. To the 50 ml sample, add 1.0 ml of sulfanilamide solution from an automatic pipette, mix and allow the reagent to react for a period greater than 2 min but not exceeding 8 min. Add 1 ml of naphthylethylenediamine solution

and mix immediately. Between 10 min and 2 hr afterwards, measure the extinction of the solution in a 1-cm cuvette against distilled water using a wavelength of 543 nm (Note b).

4. Correct the observed extinction by that of the reagent blank and calculate nitrate from the expression:

$$\mu\text{g-at N/l} = (\text{corrected extinction} \times F) - 0.95C$$

where C is the concentration of nitrite in the sample in μg-at N/l (Note c).

G. Determination of blank

Carry out the procedure in Section F, above, using 100 ml of dilute ammonium chloride instead of the seawater sample. Measure the extinction using the same cuvette as is used for the seawater samples and subtract the value for each column from the sample values for each column (Note d).

H. Calibration

Because of a small salt effect, standard nitrate solutions should be made up in synthetic seawater or a low nitrate seawater sample should be "spiked" with a standard amount of nitrate.

1. Synthetic seawater

Dissolve 310 g of analytical reagent quality sodium chloride (NaCl), 100 g of reagent quality magnesium sulfate ($MgSO_4 \cdot 7H_2O$) and 0.5 g of sodium bicarbonate ($NaHCO_3 \cdot H_2O$) in 10 l of distilled water.

2. Standard nitrate solution

Dissolve 1.02 g of analytical reagent quality potassium nitrate, KNO_3, in 1000 ml of distilled water. The solution is generally stable in the absence of evaporation.

Dilute 4.00 ml of this solution to 2000 ml with synthetic seawater. This solution should be stored in a dark bottle and prepared fresh before use.

$$\text{Concentration} \equiv 20 \ \mu\text{g-at N/l}$$

3. Procedure

Add about 100 ml of the dilute standard nitrate solution to a 125-ml Erlenmeyer flask and carry out the procedure as described in Section F. Take the extinction for each individual column (Note d); then the factor F is

$$F = \frac{20}{E_S}$$

where E_S is the extinction of the standard, corrected for the blank.

Notes

(a) Column dimensions can be scaled down proportionally and smaller seawater samples used under Section D, as required by users. Smaller dimensions are shown on Fig. 1 in brackets.

(b) For extinction values of > 1.0 or < 0.1, use an appropriate cuvette cell length (i.e., 0.5 cm or 10 cm, respectively) and adjust the factor (Section H) where appropriate.

(c) In most samples of seawater, the level of nitrite will be insignificant and the correction can be largely ignored. However, in some cases, particularly with respect to depth profiles where a nitrite maximum is expected, a correction should be employed. The factor of 0.95 allows for an approximate 5% loss of nitrite on the column compared with the direct determination as described in Method 1.2.

(d) For the blank and standard values, the extinctions obtained should be applied to individual cadmium columns and not averaged. Each column may have small consistent differences which are allowed for only if the blank and standard are applied on an individual basis.

1.2. Determination of Nitrite

Introduction

The method has been taken from the procedure described by Bendschneider and Robinson (*J. Mar. Res.*, **11**: 87, 1952).

Method

A. Capabilities

> *Range: 0.01–2.5 µg-at/l*
> *Precision at the 1 µg-at/l level:*
> The correct values lies in the range, mean of n determinations $\pm 0.03/n^{\frac{1}{2}}$ µg-at/l.

B. Outline of method

The nitrite in seawater is allowed to react with sulfanilamide in an acid solution. The resulting diazo compound is reacted with N-(1-naphthyl)-ethylenediamine and forms a highly colored azo dye.

C. Special apparatus and equipment

> 125-ml Erlenmeyer flasks
> 50-ml measuring cylinder

D. Sampling procedure and storage

Rinse glassware with samples before using; measure 50 ml of seawater from a measuring cylinder into a 125-ml Erlenmeyer flask; analyze within a few hours and preferably immediately.

E. Special reagents

1. *Sulfanilamide solution*
   Prepare as described in Method 1.1.
2. *N-(1-naphthyl)-ethylenediamine dihydrochloride solution*
   Prepare as described in Method 1.1.

F. Experimental procedure

1. Add 1.0 ml of sulfanilamide solution from an automatic pipette to each 50-ml sample, mix, and allow the reagent to react for more than 2 min but less than 10 min to assure a complete reaction.
2. Add 1.0 ml of naphthylethylenediamine reagent and mix immediately. Between 10 min and 2 hr afterwards, measure the extinction of the solution in a 10-cm cuvette at a wavelength of 543 nm.
3. Correct the measured extinction for the reagent (and turbidity, if necessary) blank and calculate the nitrite concentration as:

$$\mu\text{g-at N/l} = \text{corrected extinction} \times F$$

where *F* is determined as in Section H below.

G. Determination of blank

Carry out the method exactly as described in Section F, 1–2, using distilled water in place of seawater.

H. Calibration

1. Standard nitrite

Anhydrous, analytical grade sodium nitrite, $NaNO_2$, should be dried at 110°C for 1 hr and 0.345 g dissolved in 1000 ml of distilled water; store in a dark bottle with 1 ml of chloroform as a preservative. The solution is stable for several months.

$$1 \text{ ml} \equiv 5 \text{ } \mu\text{g-at N}$$

Dilute 10.0 ml of this solution to 1000 ml with distilled water and use the same day.

2. Procedure

Prepare three standards by pipetting 2.00 ml of dilute standard into each of 3 graduated 50-ml cylinders; make to volume of 50 ml with distilled water, mix and transfer to each of 3 Erlenmeyer flasks. Carry out steps 1 and 2 in Section F; correct the extinction for the reagent blank and calculate F from the expression:

$$F = \frac{2.00}{E_S}$$

where E_S is the mean extinction of 3 standards, corrected for the blank. The value of F is approximately 2 under conditions described above.

1.3. Determination of Ammonia (Oxidation Method)

Introduction

A sensitive photometric determination of ammonia in seawater is based upon the oxidation reaction with hypochlorite in an alkaline medium. The method has been developed by Richards and Kletch (*Sugawara Festival Volume*, Maruzen Co., Tokyo, pp. 65–81, 1961). Several investigators have improved the original method to give the best conditions: for instance, the shortening of oxidation times by the addition of a small amount of bromide (Truesdale, *Analyst*, **96**: 584, 1971); the addition of excess bromide as a catalyst (Matsunaga and Nishimura, *Anal. Chim. Acta*, **73**: 204, 1974); the critical examination of the concentrations of bromide and alkali (Shinagawa and Tsunogai, *Bull. Fac. Fish., Hokkaido Univ.*, **29**: 173, 1978); and the application for an auto-analyzer (Strickland and Parsons, *Bull. Fish. Res. Bd., Canada*, **167**, 2nd ed., 133, 1972; Le Corre and Tréguer, *J. Cons. Int. Explor. Mer*, **38**: 147, 1978). The method described here is basically taken from the procedure of Matsunaga and Nishimura (*loc. cit.*). Some improvements are carried out to give an accurate and sensitive condition and to introduce automatic analysis.

Method

A. Capabilities

Range: 0.04–10 μg-at N/l

1. Precision at the 10 μg-at N/l level:

The correct value lies in the range, mean of n determinations $\pm 0.05/n^{\frac{1}{2}}$ μg-at N/l

2. Precision at the 0.25 μg-at N/l level:

The correct value lies in the range, mean of n determinations $\pm 0.08/n^{\frac{1}{2}}$ μg-at N/l

B. Outline of method

Ammonia in seawater samples is oxidized to nitrite with hypochlorite in alkali using a large excess of potassium bromide as a catalyst. The precipitation of metal hydroxide in saline water in an alkaline medium is prevented by the addition of a complexing reagent prior to the oxidation step. Nitrite produced from the oxidation of ammonia is determined according to Method 1.2, Section F.

C. Special apparatus and equipment

Tightly stoppered 100-ml capacity Erlenmeyer flasks
Automatic pipettes with disposable tips
Constant temperature bath

D. Sampling procedure and storage

Sampling and sample storage are carried out in a similar manner to that used for other nutrients. However, the analysis of ammonia should be commenced as soon as possible, preferably within 1–2 hr after sampling. If longer storage periods are necessary, the samples should be stored in a freezer at $< -20°C$ immediately after sampling. There are indications that, even with refrigeration, losses or gains may be significant after more than a few days. It is also important to prevent the contamination of ammonia derived from the atmosphere in the laboratory.

E. Special reagents

1. Complexing reagent

Dissolve 110 g of analytical reagent quality sodium citrate, $C_6H_5O_7Na_3·2H_2O$ and 105 g of analytical grade sodium potassium tartrate, $C_4H_4O_6KNa·4H_2O$ in 1000 ml of de-ionized water. The solution is stable in a tightly stoppered bottle for many months.

2. Alkaline potassium bromide solution

Dissolve 175 g of analytical grade potassium bromide and 250 g of analytical quality sodium hydroxide in 1000 ml of de-ionized water. This solution should be stable for many months.

3. Sodium hypochlorite solution

Use a solution of commercial hypochlorite (e.g., Chlorox) which should be about 1.5N. The solution decomposes slowly and its strength should be checked periodically.

Dilute 3.5 ml of *ca.* 1.5N sodium hypochlorite solution to 100 ml with de-ionized water. The solution is stable for at least 2 hr.

4. Sodium arsenite solution

Dissolve 10 g of analytical grade sodium meta-arsenite (Na_2AsO_2) in 1000 ml of de-ionized water. This solution is stable indefinitely.

5. 8.5N hydrochloric acid

Dilute 500 ml of analytical grade concentrated hydrochloric acid (*ca.* 12N) with de-ionized water to make 8.5N HCl (i.e., dilute to 705 ml).

6. Sulfanilamide solution

Dissolve 5 g of sulfanilamide in 500 ml of 8.5N hydrochloric acid. The solution is stable for many months.

7. *N-(1-naphthyl)-ethylenediamine dihydrochloride solution*

Dissolve 0.5 g of the dihydrochloride in 500 ml of de-ionized water. Store the solution in an amber bottle. It is stable for a month.

F. *Experimental procedure (Note a)*

1. Add 50 ml of sample to an Erlenmeyer flask from a 50-ml measuring cylinder. Add 2 ml of complexing reagent from a pipette and swirl the solution (Note b).

2. Add 2 ml of alkaline potassium bromide solution from a pipette, swirl the solution, and allow the flask to stand at a temperature between 35° and 45°C (Note c).

3. Add 2 ml of 0.05N sodium hypochlorite solution, swirl vigorously, and allow the flask to stand for 2 min at 35–45°C (Note d).

4. Add 2 ml of 1% sodium meta-arsenite solution, swirl the solution, and allow the flask to stand at room temperature (20–25°C) for 2 min (Note e).

5. Add 2 ml of sulfanilamide solution and swirl the solution. Allow the reagent to react for a period greater than 2 min but not exceeding 8 min. Add 2.0 ml of N-(1-naphthyl)-ethylenediamine solution and mix immediately. Between 10 min and 2 hr afterwards, measure the extinction of the solution in a 1- or 5-cm cell against de-ionized water at a wavelength of 543 nm (Note f).

6. Correct the measured extinction by that of the reagent blank (Section G) and calculate the ammonium nitrogen concentration from the expression:

$$\mu\text{g-at N/l} = F \times \left(E - \frac{0.838 \times c}{F'} \right)$$

where E is the corrected extinction from which the reagent blank has been subtracted, F' is the nitrite factor (Method 1.2, Section H), and c is the concentration of nitrite in μg-at/l. The numerals 0.838 represent the dilution factor between ammonia and nitrite. F is the factor obtained as described in Section H below (Note g).

G. *Determination of blank*

Carry out the method exactly as described in Section F, steps 1–6, using 50 ml of ammonium-free seawater (Note a–4) for the seawater blank value and 50 ml of de-ionized water for the reagent blank.

H. *Calibration*

1. Carry out the calibration with filtered seawater in which the concentration of ammonia has been reduced by boiling.

Dissolve 0.100 g of analytical grade ammonium sulfate in 1000 ml of de-ionized water. Add 1 ml of chloroform as a preservative and store in a refrigerator. The solution is stable for many months if well stoppered.

$$1 \text{ ml} \equiv 1.5 \ \mu\text{g-at N}$$

Pipette 1 ml of this solution into a 500-ml measuring flask and make to volume with ammonium-free seawater. The concentration is equivalent to 3 μg-at N/l. This secondary standard solution should be made up fresh before use.

2. Add 50 ml of the secondary standard to an Erlenmeyer flask from a 50-ml measuring cylinder. Carry out the procedure as described in Section F, steps 1–6. Calculate the factor F from the expression:

$$F = \frac{3.0}{E_s - E_b}$$

where E_s is the mean extinction of the three standards and E_b is the mean extinction of three seawater blanks as measured in Section G. The values for F should be near 7.4 in a 5-cm cell.

Notes

(a) General precautions against contamination by ammonia:

(1) De-ionized water should be passed through a strong cation exchange resin (e.g., Dowex 50 × 8, hydrogen form) immediately before use.

(2) All reagents should be stored in well stoppered containers which have been washed with dilute hydrochloric acid.

(3) Ammonium hydroxide should not be used in the laboratory during this experiment.

(4) Ammonium-free seawater is desirable in the calibration. If it is not available, then 1 l of seawater should be boiled after addition of 5 ml of 1N NaOH and the volume reduced until it is about 0.7 l. The volume is replaced to 1 l with de-ionized water after neutralization with equivalent HCl and is filtered through a glass fiber filter.

(5) The complexing reagent solution frequently has high blank values. It is necessary to remove the ammonium contamination by boiling the solution in alkali (pH 10 ~ 11).

(b) The complexing reagent is added in order to obtain a constant rate of oxidation by preventing the adsorption of ammonia into the precipitates which can be produced from a seawater sample under alkali conditions. This procedure is especially important if an automatic analysis is used, since it removes the adherence of precipitates on the inner surfaces of thin glass or plastic tubes.

(c) The time required for maximum oxidation is reduced by the use of a large excess of bromide as catalyst. In the present method, the concentration of bromide is 2.5 times that in the original paper (Matsunaga and Nishimura, 1974) according to Shinagawa and Tsunogai (1978).

(d) An optimum condition for oxidation is given in the concentration ranges of 0.025 to 0.050 normal hypochlorite at a temperature of 35° to 45° for 2 min. Amino acids in a sample do not interfere because of the shortened oxidation time. However, if longer oxidation times are used, they will give an apparently higher value owing to the oxidation of amino acids. Hydroxyl-amine is completely oxidized to nitrite and gives a positive error, although the compound is negligible in oxygenated seawater at pH 8.

(e) If fresh sodium meta-arsenite is not available, arsenic trioxide (As_2O_3) could be used. Dissolve 5 g of analytical quality arsenic trioxide in 200 ml of de-ionized water and *ca.* 1.5 g of analytical grade sodium hydroxide pellets. Dilute the solution to 500 ml with de-ionized water.

(f) The extinction of the resultant solution is measured using 1- or 5-cm cells. If the concentration is less than 3 μg-at/l, the extinction should be measured with a 5-cm cell.

(g) Any nitrite initially present in the sample is unchanged by the analytical procedure, so a correction for its presence can be made in the manner shown. In this calculation, allowance is made for the fact that the sample is diluted from 52 to 62 ml by reagents before extinction measurements are made and that only a fraction of the ammonia is converted to nitrite. This correction can be quite significant in water containing a relatively large amount of nitrite but little ammonia.

1.4. Determination of Ammonia (Alternative Method)

Introduction

The following method is suggested as an alternate method. It is less sensitive than the previous method (Method 1.3), but it is specific for ammonia and may be found easier to apply. The method is taken from several authors (Riley, *Anal. Chim. Acta*, **9**: 575, 1953; Emmet, *Naval Ship-Res. Div. Cent. Rept.* 2570, 1968; Solorzano, *Limnol. Oceanogr.*, **14**: 799, 1969).

Method

A. *Capabilities*

> *Range: 0.1 to 10 μg-at/l*
> *Precision at the 1 μg-at/l level:*
> The correct value lies in the range, mean of n determinations $\pm 0.1/n^{\frac{1}{2}}$ μg-at/l

B. *Outline of method*

Seawater is treated in an alkaline citrate medium with sodium hypochlorite and phenol in the presence of sodium nitroprusside which acts as a catalyzer. The blue indophenol color formed with ammonia is measured spectrophotometrically.

C. *Special apparatus and equipment*

125-ml Erlenmeyer flasks. These must be cleaned with dichromate acid and rinsed thoroughly with distilled water.

Automatic pipettes to dispense reagents.

D. *Sampling procedure and storage*

Temporary storage of seawater prior to analysis appears satisfactory in glass or polyethylene bottles, but analysis should not be delayed for more than 1–2 h at the most. If analyses cannot be performed in this time period, samples should either be frozen at $-15°C$ or stored unfrozen in the presence of 2 ml of phenol solution/50 ml of sample (Reagent 2, Section E). Samples may be stored in either manner for up to two weeks (Degobbis, *Limnol. Oceanogr.*, **18**: 146, 1973).

E. *Special reagents*

1. *De-ionized water*

Distilled water is passed through a cation exchange column in the hydrogen form (30 cm long, 1–2 cm wide). This water should be prepared fresh for use.

2. *Phenol solution*

Dissolve 20 g of analytical grade phenol in 200 ml of 95% v/v ethyl alcohol.

3. Sodium nitroprusside solution

Dissolve 1.0 g of sodium nitroprusside, $Na_2[Fe(CN)_5NO]\cdot 2H_2O$, in 200 ml of de-ionized water. Store in a dark glass bottle; the solution is stable for at least a month.

4. Alkaline reagent

Dissolve 100 g of sodium citrate and 5 g of sodium hydroxide in 500 ml of de-ionized water. The solution is stable indefinitely.

5. Sodium hypochlorite solution

Use commercially available hypochlorite (e.g., "Chlorox") which should be about 1.5N. The solution decomposes slowly and should be checked periodically (Note a).

6. Oxidizing solution

Mix 100 ml of reagent 4 and 25 ml of reagent 5. Keep stoppered while not in use and prepare fresh every day.

F. Experimental procedure (Note b)

1. Add 50 ml of seawater to an Erlenmeyer flask from a 50-ml measuring cylinder. Add 2 ml of phenol solution, swirl to mix, and then add in sequence 2 ml of nitroprusside and 5 ml of oxidizing solution (Note c); mix after each addition by swirling the flasks.
2. Allow the flasks to stand at room temperature (20–27°C) for 1 hr. The top of the flask should be covered with "parafilm" or "saranwrap" during this period. The color is stable for *ca.* 24 hr after the reaction period (Note d).
3. Read the extinction at 640 nm in a spectrophotometer using a 10-cm cell length.
4. Correct the measured extinction for the reagent blank and calculate ammonia-nitrogen from the expression:

$$\mu\text{g-at N/l} = F \times E$$

where E is the corrected extinction and F is the factor as determined in Section H.

G. Determination of blank

Carry out the method exactly as described in Sections F1 to F3 above using

50 ml of de-ionized water. Blank extinctions should not exceed about 0.075 on a 10-cm cell.

H. Calibration

1. Prepare 50 ml of standard ammonia solution as described in Method 1.3, Section H1.

2. Carry out the determination above, Sections F1 to F3. Correct the extinction for the reagent blank and calculate F as:

$$F = \frac{3.0}{E_S}$$

where E_S is the corrected extinction. The value of F should be about 6.5 and standardization should be run in triplicate.

Notes

(a) To check on the strength of hypochlorite, dissolve 12.5 g of sodium thiosulfate ($Na_2S_2O_3 \cdot 5H_2O$) in 500 ml of water. Add a few crystals (*ca.* 2 g) of potassium iodide (KI) to about 50 ml of water in a small flask and pipette in 1.0 ml of hypochlorite solution. Add 5–10 drops of concentrated hydrochloric acid (HCl) and titrate the liberated iodine with the thiosulfate solution until no yellow color remains. Discard the hypochlorite when less than 12 ml of thiosulfate is used.

(b) Great care is necessary in this method to reduce contamination from external sources. Plastic gloves should be worn over the hands and no use of ammonia for other analyses should be allowed in the same laboratory.

(c) Some seawater samples may require more than 5 ml of oxidizing reagent. However, the pH after addition of this reagent must not exceed *ca.* 9.8.

(d) Serious overdevelopment of the blue color may occur in high light intensities. This requires flasks to be wrapped in aluminum foil (see Gravitz and Gleye, *Limnol. Oceanogr.*, **20**: 1015, 1975).

1.5. Determination of Urea

Introduction

There are two different methods for determination of urea in seawater. One is a urease method which has been developed by McCarthy (*Limnol.*

Oceanogr., **15**: 309, 1970) using the ammonia method described by Solorzano (*Limnol. Oceanogr.*, **14**: 799, 1969). The other is a direct chemical method given by Newell (*J. mar. biol. Ass U.K.*, **47**: 271, 1967). The latter method has been modified for application to seawater by Koroleff (In: *Methods of Seawater Analysis*, edited by Grasshoff, 1976). We recommend the urease method because the method is specific for urea and more sensitive than the direct chemical method. The method is basically taken from McCarthy (*loc. cit.*) with a few modifications.

Method

A. *Capabilities*

Range: 0.05–5 µg-at urea-N/l
Precision at the 3 µg-at/l level:
   The correct value lies in the range, mean of *n* determinations $\pm\,0.1/n^{\frac{1}{2}}$ µg-at/l

B. *Outline of method*

   Urea in seawater is decomposed to ammonia in the presence of a specific enzyme, urease, which acts as a catalyst for the reaction. The resulting ammonia is determined by spectrophotometry using an alkaline phenol method.

C. *Special apparatus and equipment*

   125-ml Erlenmeyer flasks
   Water bath at 50°C
   Cellulose dialyzer tubing (approximately 20 cm long by 3 cm diameter)

D. *Sampling procedure and storage*

   Water samples are collected with PVC bottles together with the samples for nutrients and organic matter. It is especially important to take precaution against contamination from the ship's waste water in the collection of samples.
   Determination of urea should be carried out immediately after sampling but, if this is impossible, the samples should be stored at temperatures below $-20°C$.

E. Special reagents

1. De-ionized water

Remove the ammonia from distilled water by passing it through a cation exchange column, as described in ammonium determination (Method 1.3).

2. Ethylene diamine tetra acetate (EDTA)

Dissolve 10 g of the disodium salt of EDTA in 900 ml of distilled water. Adjust the pH to 6.5 with a little (*ca.* 5% w/v) sodium hydroxide solution and dilute to 1 l.

3. Cleland's reagent

Dissolve 0.2 g of Cleland's reagent (dithrothreitol) in 100 ml of distilled water. Store in a glass bottle and keep frozen in a deep-freezer when not in use.

4. Concentrated enzyme solution

Crude urease (preferably Worthington (URC) lyophilized) will be contaminated with ammonium salts and it is best purified by dialysis.

Dissolve 0.25 g of enzyme in 45 ml of EDTA solution and pour it into a dialysis tube in a 1-l beaker containing about 800 ml of EDTA chilled to 5°C. Place the beaker in a refrigerator (3–6°C) and stir the solution changing the EDTA in the beaker every half day for 3 days. Remove the dialyzed enzyme solution into a glass bottle and add 50 ml of analytical reagent quality glycerol and 5 ml of Cleland's reagent. Store the mixture at 5°C. The preparation is stable for several months if kept cold.

5. Dilute enzyme solution

Dilute 10 ml of concentrated enzyme solution to 100 ml with de-ionized water. Prepare fresh each day.

6. Phenol solution

See Method 1.4, Section E2.

7. Sodium nitroprusside solution

See Method 1.4, Section E3.

8. Alkaline reagent

See Method 1.4, Section E4.

9. Sodium hypochlorite solution

See Method 1.4, Section E5.

10. Oxidizing solution

Mix 100 ml of reagent 8 with 25 ml of reagent 9. Prepare fresh every day.

F. Experimental procedure

1. Add 50 ml of sample to each of four 125-ml Erlenmeyer flasks using a 50-ml measuring cylinder. Add 5.0 ml of dilute enzyme solution from a pipette to two (only) of the flasks and mix the contents. Cover the mouths of all four flasks with aluminum foil (Note a).

2. Place the two flasks containing the enzyme in a water bath at 50°C for 10 min (Note b).

3. Cool the flasks to room temperature. Add 5.0 ml of dilute enzyme to the two aliquots that have not been heated (Note c).

4. Carry out the determination of ammonia on all four flasks exactly as described in Method 1.4 (Note d).

5. Read the extinction of all four flasks in a spectrophotometer against water in 10-cm cells using a wavelength of 640 nm.

6. Calculate the urea nitrogen concentration from the expression:

$$\mu\text{g-at N/l} = F \times (E_S - E_B)$$

where E_S is the mean extinction of the solutions in the two flasks heated with urease and E_B is the mean extinction of the solutions in the unheated flasks (Note e). F is a factor determined as described later in Section H.

G. Determination of blank

1. Reagent blank

There is no reagent blank *per se* for this method as individual blanks are carried out for each sample along with the main determinations.

H. Calibration

1. Standard urea solution

Dissolve 0.09 g of analytical reagent quality urea in 1000 ml of distilled water. Add a few milliliters of chloroform. The solution is stable at room temperature, if not contaminated, but it is best stored at 5°C.

$$1 \text{ ml} \equiv 3 \text{ } \mu\text{g-at urea N}$$

2. Dilute standard

Add 1.00 ml of the above solution to 1 l of low-ammonia seawater. The concentration is then 3.0 μg-at urea N/l.

3. Procedure

Measure out three 50-ml portions of the dilute standard into clean Erlenmeyer flasks. Add to three more flasks the same seawater that was used to prepare the dilute standards but with no added urea. Carry out the determination exactly as described in Section F, 1–6.

Calculate the factor F from the expression:

$$F = \frac{3.0}{(E_S - E_B)}$$

where E_S and E_B are the mean extinctions of the triplicate standards and standard blanks, respectively. The value for F should be near to 6.5 and reasonably constant.

Notes

(a) Foil or screw stoppers must be used to close the mouth of each flask to prevent variable atmospheric contamination from ammonia.

(b) Depending on the enzyme preparation, some experimenters have found a better hydrolytic efficiency when the enzyme is incubated at 80°C. The optimum pH for hydrolysis is in the neutral range, therefore it is not required to add buffer to normal seawater.

(c) The unheated aliquots act as blanks for the ammonia in the seawater sample and for any residual ammonia in the enzyme preparation. The latter should not exceed the equivalent of about 0.15 μg-at N/l.

(d) The determination of ammonia should be done without delay (0.5–1 min) after adding the enzyme to the two cold blanks, so that no hydrolysis of urea takes place. Hydrolysis is very slow at room temperature (25°C) but not negligible. It is stopped completely when the pH is elevated after adding the ammonia reagents.

(e) The blank determination is not a reliable estimate of any ammonia in seawater because of possible reagent contamination; however, ammonia determinations are conveniently carried out at the same time as urea determinations.

1.6. Determination of Phosphate

Introduction

The procedure given below is taken from Murphy and Riley (*Anal. Chim. Acta*, **27**: 31, 1962).

Method

A. Capabilities (Note a)

Range: 0.03 to 5 μg-at/l
Precision at the 3 μg-at/l level:
  The correct value lies in the range, mean of n determinations $\pm 0.03/n^{\frac{1}{2}}$ μg-at/l

B. Outline of method

The seawater sample is allowed to react with a composite reagent containing molybdic acid, ascorbic acid and trivalent antimony. The resulting complex is reduced to give a blue solution which is measured at 885 nm.

C. Special apparatus and equipment

130-ml capacity screw-capped polyethylene bottles marked on the side with 100 ml volume with black tape, or Erlenmeyer 125-ml graduated flasks.

D. Sampling procedure and storage

When possible, samples should be analyzed as soon as they are collected. Polyethylene bottles should be rinsed twice before filling; samples may be stored at $-20°C$, but there is conflicting evidence on the effect of storage (cf., Hassenteufel *et al.*, *Limnol. Oceanogr.*, **8**: 152, 1963) and any effect should be checked by individuals using their own storage bottles.

E. Special reagents

1. Ammonium molybdate solution

Dissolve 15 g of analytical reagent grade ammonium paramolybdate $(NH_4)_6Mo_7O_{24} \cdot 4H_2O$ in 500 ml of distilled water. Store in plastic bottle away from direct sunlight. The solution is stable (Note b).

2. Sulfuric acid solution

Add 140 ml of concentrated (sp. gr. 1.82) analytical reagent quality sulfuric acid to 900 ml of distilled water. Allow the solution to cool and store it in a glass bottle.

3. Ascorbic acid solution

Dissolve 27 g of ascorbic acid in 500 ml of distilled water. Store the solution in a plastic bottle frozen solid in the freezer. The solution is stable for many months but should not be kept at room temperature for more than one week.

4. Potassium antimonyl-tartrate solution

Dissolve 0.34 g of potassium antimonyl-tartrate (tartar emetic) in 250 ml of water, warming if necessary. Store in a glass or plastic bottle. The solution is stable for many months.

5. Mixed reagent

Mix together 100 ml ammonium molybdate, 250 ml sulfuric acid, 100 ml ascorbic acid, and 50 ml of potassium antimonyl-tartrate solutions. Prepare this reagent when needed and discard any excess. Do not store; the quantity is suitable for about 50 samples.

F. Experimental procedure

1. Warm the samples to room temperature (15–30°C). Measure the turbidity of a sample at 885 nm (Note c); if this value is greater than 0.01, a correction should be applied to the final extinction value (Step 4).

2. To a 100 ml sample, add 10 ml of mixed reagent using a syringe-type pipette and mix at once.

3. After 5 min, and preferably within the first 2–3 h (Note d), measure the extinction in a 10-cm cell against distilled water at 885 nm.

4. Correct the extinction with the reagent blank (and turbidity blank if necessary) and calculate the phosphate concentration as:

$$\mu\text{g-at P}/l = \text{corrected extinction} \times F$$

where F is the factor as described below.

G. *Determination of blank*

Use distilled water in place of a sample and carry out steps 1–4 above to obtain the extinction of the reagent blank. Reagent blanks should be less than 0.02 on a 10-cm cell.

H. *Calibration*

Dissolve 0.816 g of anhydrous potassium dihydrogen phosphate, KH_2PO_4, in 1 l of distilled water. Store in a dark bottle with 1 ml of chloroform; the solution is stable for many months.

Dilute 10 ml of the standard to 1 l with distilled water. Pipette 5 ml of dilute standard into each of 3 Erlenmeyer flasks and make up to 100 ml with distilled water from a graduated cylinder. Carry out the procedure in steps 2 to 4, Section F above. Calculate the factor, F, as:

$$F = \frac{3.00}{E_S - E_B}$$

where E_S is the average extinction of three standards and E_B is the average extinction of the reagent blank. The value of $F_{10\,cm}$ should be about 5.

Notes

(a) A method for determining very low levels of phosphate, particularly such as are encountered in some tropical waters, is given by Stephens (*Limnol. Oceanogr.*, **8**: 361, 1963). The method is the same as that described here except that the blue color produced is concentrated by extraction with iso-butanol and measured at 690 nm. The method results in an approximate 5-fold increase in sensitivity.

(b) If high blank values are experienced, the ammonium molybdate should be made up again and tested.

(c) This should generally be appreciably less than 0.01 on a 10-cm cell; under heavy bloom conditions or in the presence of silt, it may be considerably greater. Samples should either be filtered through a glass fiber filter or a correction applied for each sample with a turbidity value greater than 0.01.

(d) The extinction reaches a maximum in 5 min and is stable for several hours. However, a small increase may occur if samples are left for half a day or longer.

Addendum

Using the same reagents, total phosphorus (i.e., organic soluble and particulate phosphorus together with inorganic phosphate) can be obtained by difference subtraction of filtered and unfiltered seawater samples, after treatment with perchloric acid (Hansen and Robinson, *J. Mar. Res.*, **12**: 31, 1953) or persulfate (Menzel and Corwin, *Limnol. Oceanogr.*, **10**: 280, 1965). Another popular method, which can be used for either organic phosphorus or nitrogen, is to carry out the oxygenation under high intensity ultraviolet light (Armstrong, Williams and Strickland, *Nature*, **211**: 481, 1966).

1.7. Determination of Silicate

Introduction

The method described here is basically taken from Mullin and Riley (*Anal. Chim. Acta*, **12**: 162, 1955). The method measures "reactive" silicate and does not measure polymerized silicate; however, some of the "reactive" silicate may not be biologically available. Paasche (*Mar. Biol.*, **19**: 262, 1973) has shown that between 0.3 and 1.3 μg-at Si/l, which could be measured chemically in seawater, could not be taken up by diatoms. This would be a small fraction of the amount of silicate in many seawater samples and generally could be ignored.

Method

A. Capabilities

Range: 0.1 to 140 μg-at/l
Precision at the 10 μg-at/l level:
The correct value lies in the range, mean of n determinations $\pm 0.25/n^{\frac{1}{2}}$ μg-at/l

B. Outline of method

The seawater sample is allowed to react with molybdate under conditions which result in the formation of silicomolybdate, phosphomolybdate and arsenomolybdate complexes. A reducing solution, containing metol and oxalic acid, is then added which reduces the silicomolybdate complex to give a blue color and simultaneously decomposes any phosphomolybdate or arsenomolybdate. The resulting extinction is measured using a 1- or 10-cm cuvette, depending on concentrations encountered.

C. Special apparatus and equipment

50-ml capacity, stoppered, graduated glass cylinders, one for each seawater sample. Cylinders should be cleaned initially with chromic-sulfuric acid and kept for silicate analysis, rinsing after use with distilled water.

D. Sampling procedure and storage

Samples should be collected in screw-cap polyethylene bottles (*ca.* 125-ml). Samples may be frozen at $-20°C$ but some loss of silicate at very high concentrations (e.g., deep samples > 50 μg-at/l) may occur; also, in high diatom blooms, some regeneration of soluble silicate can occur if plankton is not removed by filtration. Therefore, it is advisable to carry out the determination of silicate when the samples are first collected.

E. Special reagents

1. Molybdate reagent

Dissolve 4.0 g of analytical reagent quality ammonium paramolybdate, $(NH_4)_6Mo_7O_{24}\cdot4H_2O$, in about 300 ml of distilled water. Add 12 ml of concentrated hydrochloric acid (12N), mix and make to volume of 500 ml with distilled water. Store the solution in a polyethylene bottle and keep out of direct sunlight.

2. Metol-sulfite solution

Dissolve 6 g of anhydrous sodium sulfite, Na_2SO_3, in 500 ml of distilled water and add 10 g of metol (p-methylaminophenol sulfate). When the metol has dissolved, filter the solution through a No. 1 Whatman filter paper and store it in a clean glass bottle which is tightly stoppered. This solution may deteriorate quite rapidly and erratically and should be prepared fresh at least once a month.

3. Oxalic acid solution

Prepare saturated oxalic acid solution by shaking 50 g of analytical reagent quality oxalic acid dihydrate, $(COOH)_2 \cdot 2H_2O$, with 500 ml of distilled water. Decant the solution from the crystals for use; the solution may be stored in a glass bottle and is stable indefinitely.

4. Sulfuric acid solution 50% v/v

Pour 250 ml of concentrated sulfuric acid (sp. gr. 1.82) into 250 ml of distilled water. Cool to room temperature and make the volume to 500 ml with a little extra water.

5. Reducing reagent

Mix 100 ml of metol-sulfite solution with 60 ml of oxalic acid solution. Add slowly, with mixing, 60 ml of the 50% sulfuric acid solution and make the mixture up to 300 ml with distilled water. The solution should be prepared each time for immediate use.

F. Experimental procedure

1. Samples should be at room temperature (18–25°C). Add 10 ml of molybdate solution to a dry 50-ml graduated cylinder fitted with a glass stopper. Pipette 25 ml of the seawater sample into the cylinder, stopper, and mix by inverting; allow to stand for 10 min, but not for more than 30 min.

2. Add the reducing reagent rapidly (it can be dispensed from a wash bottle) to make 50 ml and mix immediately.

3. Allow the solution to stand for 2–3 hr (Note a) to complete the reduction. Measure the extinction at 810 nm using a 1-cm cell for concentrations > 15 μg-at/l and a 10-cm cell for concentrations of < 15 μg-at/l.

4. Correct the measured extinction for the blank (1- or 10-cm cell length) and calculate reactive silicate as:

$$\mu\text{g-at Si/l} = \text{corrected extinction} \times F$$

where F is the factor for 1- or 10-cm cell length as defined below.

G. Determination of blank

Use distilled water, which has been collected in a polyethylene container, in place of seawater and carry out steps 1 to 4 in Section F.

H. Calibration

1. Standard silicate solution

Weigh 0.960 g (deliberately a slight excess over the theoretical) silicofluoride, $Na_2 SiF_6$, and dissolve in 100 ml of distilled water. Dilute to exactly 1000 ml, mix, and immediately transfer the solution to a polyethylene container for storage. The solution is stable and consists of

$$1 \text{ ml} \equiv 5 \text{ } \mu\text{g-at Si}$$

Dilute 10 ml of this solution to 500 ml with synthetic seawater (25 g NaCl plus 8 g $MgSO_4 \cdot 7H_2O/l$ of distilled water).

2. Procedure

Carry out the silicon determination in Section F using 25 ml of the standard silicate solution in synthetic seawater. Measure the extinction on a 1-cm cell after 3 hr. Then,

$$F_{1 \text{ cm}} = \frac{100}{E_S - E_B}$$

where E_S is the extinction of the standard and E_B is the extinction of the blank. The value of $F_{1 \text{ cm}}$ should be about 100; $F_{10 \text{ cm}}$ will be $F_{1 \text{ cm}} \times 10^{-1}$ or about 10.

Notes

(a) The reaction is generally complete in 1 hr except when high silicates are encountered. Solutions should be measured within 6 hr.

1.8. Bioassay of Seawater for Vitamins: Vitamin B_{12} (Cyanocobalamin)

Introduction

For certain organic and inorganic constituents of seawater, the quantities present (in $\mu\mu$g/ml, or picog/ml) may be best determined by a bioassay rather than a chemical technique. The example of such a technique given here is for vitamin B_{12} analysis and the method is taken from Gold (*Limnol. Oceanogr.*, **9**: 343, 1964) with some modifications suggested by Carlucci and Silbernagel (*Can. J. Microbiol.*, **12**: 175, 1966). While both authors use the diatom, *Cyclotella nana*, for measuring uptake, in the procedure described below, the dinoflagellate, *Amphidinium carteri*, is used here since this avoids the step of having to add neutralized silicate to the culture medium.

The assay of other growth factors follows a very similar procedure to that described below if the medium is modified to be limiting to the substance for which the analysis is required and vitamin B_{12} is added in excess. Examples of such analyses are given for biotin (Carlucci and Silbernagel, *Can. J. Microbiol.*, **13**: 979, 1967) and for thiamine (Carlucci and Silbernagel, *Can. J. Microbiol.*, **12**: 1079, 1966). The choice of organisms used in such assays depends upon a demonstrated requirement for the chemical species being assayed.

Method

A. Capabilities

Range: 0.05–3 mμg B_{12}/l (Note: the range can be extended to values > 3 mμg B_{12}/l by dilution of seawater samples with vitamin B_{12}-free seawater.)
Precision at the 1 mμg B_{12}/l level:
The correct value lies in the range, mean of *n* determinations $\pm 0.15/n^{\frac{1}{2}}$ mμg B_{12}/l (*Note*: the precision is largely determined by the presence or absence of inhibitors. The precision quoted is for replicate samples where no inhibitors were present.)

B. Outline of method

A standard curve is prepared from vitamin-free seawater and known additions of vitamin B_{12}. The response of the culture organisms to unknown concentrations of vitamin B_{12} in seawater samples is then compared to the standard curve response. Samples that are above the standard curve response may be diluted with vitamin-free seawater. It may also be possible to improve the analysis of samples falling below the standard curve response through improvement in blanks and by the use of internal standards.

C. Special apparatus and equipment

20×125-mm screw-top test tubes (cleaned with detergent, well rinsed with distilled water and dried at 250°C for 3 hr)
125-ml Erlenmeyer flasks with cotton wool plugs enclosed in cheesecloth
10, 5, 1 and 0.1-ml graduated pipettes for the preparation of media
Autoclave
Incubator at 20 ± 2°C having fluorescent lights giving an even light field of *ca.* 100 $\mu E/m^2/s$
Millipore filtering apparatus and 0.22 μ GS Millipore filters

D. Sampling procedure and storage

Samples should be filtered soon after collection. In order to avoid plugging of filters, 50-ml subsamples, from water which has been filtered for nutrient analysis through Millipore AA (0.8 μ) filters, can be filtered through Millipore GS (0.22 μ) filters into flasks under aseptic conditions (Note a). The samples are stored at $-20°C$ until analyzed.

A sample of seawater from the same general area as the unknown may be collected at the same time. This can be used to prepare vitamin-free seawater which may be necessary for dilution of samples.

E. Special reagents

1. Vitamin-free seawater

Add 10 g of Norite-A activated charcoal to 1 l of seawater; shake for 30 min and filter off the charcoal using a Whatman No. 2 filter (Note b). Filter the seawater aseptically through a GS Millipore filter and store. Prepare new vitamin-free seawater for each new group of analyses.

2. Nutrient enrichment solutions

(i) Nitrate and phosphate
Dissolve 10 g of potassium nitrate (KNO_3) and 1.4 g of potassium dihydrogen phosphate (KH_2PO_4) in 1 l of distilled water. Autoclave at 15 psi for 15 min and store aseptically in the dark.

(ii) Chelated metals
(a) Dissolve 0.08 g of cobalt chloride ($CoCl_2 \cdot 6H_2O$) and 0.08 g of copper sulfate pentahydrate ($CuSO_4 \cdot 5H_2O$) in 100 ml of distilled water. Add 1 ml of this solution to (b) below.
(b) Dissolve: 0.2 g ferric chloride ($FeCl_3 \cdot 6H_2O$)
          0.06 g zinc sulfate ($ZnSO_4 \cdot 7H_2O$)
          0.12 g manganous sulfate ($MnSO_4 \cdot H_2O$)
          0.03 g sodium molybdate ($Na_2MoO_4 \cdot 2H_2O$)
          1.2 g disodiumethylenediaminetetraacetate
in 800 ml of distilled water with 1 ml of solution (a) above. Adjust the pH to 7.5–7.8 with dilute NaOH. Dilute to 1 l and autoclave at 15 psi for 15 min. Store aseptically in the dark.

(iii) Dissolve 10 mg of thiamine hydrochloride and 10 mg of biotin in 100 ml of distilled water. Dilute 10 ml of this solution to 100 ml. Sterilize the diluted solution by passing it through a UF fritted glass filter. Store both the concentrated and diluted vitamin solutions at $-20°C$ in glass bottles; the diluted vitamin solution is conveniently dispensed into 10 ml aliquots in screw-cap test tubes for use with each batch of samples.

3. Radioactive bicarbonate

Sealed ampules containing 1 μCi ^{14}C-sodium carbonate in 1 ml of 3.5% saline are either purchased or made up and sterilized by autoclaving.

4. Algal innoculum

To each of 3 sterile 125-ml Erlenmeyer flasks, add 50 ml of vitamin-free seawater and 0.25 ml of nutrient solutions 2(i) and 2(ii) and 0.05 ml of nutrient solution 2(iii). Additions should be made aseptically and the flasks stoppered with cotton wool in cheesecloth. To one flask, add 0.05 ml of vitamin B_{12} solution (Section G); this flask is the stock solution and is used to maintain the culture. Transfer 0.5 ml of actively growing axenic *Amphidinium carteri* to the stock solution and incubate for 3 days in the light incubator. Transfer 0.5 ml of this solution to one of the vitamin B_{12}-free flasks and incubate for a further 3 to 5 days. Add 0.5 ml of this solution to the remaining flask; use this culture after 3 days for the innoculum (Note c).

F. Experimental procedure

1. Thaw samples and dispense 20 ml of sample into each of two test tubes (Note d).
2. Add to each tube 0.1 ml of nutrient solutions 2(i) and 2(ii), and 0.02 ml of 2(iii) (Note e). To the duplicate tube for each unknown sample, add 0.2 ml of dilute standard vitamin B_{12} solution to give a concentration of 1 mμg/l for the internal standard.
3. To each tube, add a 0.5 ml aliquot of the *Amphidinium carteri* innoculum. Mix the contents of each tube by swirling.
4. Place all the flasks in the incubator and incubate for at least 48 hr (Note f).
5. Add 1 ml of ^{14}C-bicarbonate solution containing 1 μCi to each tube, mix by swirling and return to the incubator for exactly 2 hr.
6. Filter the contents of each tube onto 25 mm Millipore AA filters; rinse the filters three times with filtered seawater to remove the inorganic bicarbonate and place in a scintillation tube. Add a suitable scintillation cocktail as discussed in Method 5, Section E4, and count the samples (cpm) 24 hr later.
7. Read the apparent concentration of vitamin B_{12} from the radioactive count obtained using the standard curve as prepared with each batch of samples in Section G below. Let this concentration be A in mμg B_{12}/l. Read the concentration in the same sample flask to which 1 mμg B_{12}/l was added and let this be B in mμg/l. Calculate the concentration of B_{12} in the sample as:

$$\text{m}\mu\text{g } B_{12}/\text{l} = \frac{A}{B-A} \times \frac{20}{V}$$

where V is the number of ml introduced into the test tube for analysis in Step 1, Section F (Note d).

G. Calibration

1. Standard cobalamine (vitamin B_{12}) solution

Dissolve 11.0 mg of pure crystalline vitamin B_{12} in 100 ml of distilled water (Note g). Sterilize by filtering through a UF grade fritted glass filter and store in 10 ml aliquots at $-20°C$.

2. Dilute vitamin B_{12} standard

Dilute 1.0 ml of concentrated standard solution prepared above to 100 ml. Dilute 1 ml of this second dilution to 100 ml with distilled water. Finally, dilute this solution 1 ml in 100 ml; then

$$1 \text{ ml} = 0.1 \text{ m}\mu\text{g } B_{12}$$

(0.2 ml of this solution is used as the internal standard in Step F-2.)

3. Procedure

With each series of samples being analyzed, it is necessary to construct a standard curve. The curve can be constructed from five or more concentrations, preferably carried out in duplicate.

To each of a series of test tubes, add 20 ml of vitamin-free seawater, 0.1 ml of nutrient solutions 2(i) and 2(ii), 0.02 ml of solution 2(iii) and a gradation of dilute vitamin B_{12} standard from 0.05, 0.10, 0.20, 0.30 and 0.50 ml. Carry out the procedure F, steps 3 to 6, and construct a standard curve of counts/min (cpm) per sample vs. the added vitamin B_{12} content expressed in $m\mu g$ B_{12}/l (i.e., in the series given above, 0.25, 0.5, 1.0, 1.5 and 2.5 $m\mu g$ B_{12}/l).

Notes

(a) Aseptic conditions following filtration may be difficult to achieve under field conditions. Since the incubation time is short, it is not entirely necessary to collect the GS-filtrate aseptically.

(b) With some brands of charcoal, it may be necessary to pre-wash the charcoal in 5% NaCl.

(c) Some trial and error is necessary with the innoculum, but the final number of cells should be between *ca.* 1 and 5×10^5 cells/ml.

(d) If a dilution of the sample is required, add 5 or 10 ml of sample and make up to 20 ml with vitamin B_{12}-free seawater.

(e) To avoid making several aseptic additions to each tube prior to incubation, aliquots of the three nutrient solutions can be mixed in the right proportions just prior to the analysis and dispensed as a single addition into each tube.

(f) For the ^{14}C assay, 48 hr is generally sufficient, but it may be necessary to incubate for 96 hr in order to establish a graded growth response. If, as an alternative, cell growth is measured by optical density, chlorophyll content or as the total number of cells using a Coulter counter, then the cultures can be allowed to grow for longer (e.g., 7 days).

(g) Crystalline B_{12} usually contains about 10% water and therefore the amount of vitamin is equivalent to 10 mg.

SECTION 2

Soluble Organic Material

2.1. Determination of Total Dissolved Organic Nitrogen

Introduction

The following method is taken from Solorzano and Sharp (*Limnol. Oceanogr.*, **25**: 751, 1980). The method is easier to apply than the Kjeldahl method and, at the same time, the oxidation of organic compounds is more complete than with the ultraviolet light technique (see Addendum to Method 1.6). For purposes of the analysis, the soluble fraction is defined as the organic material which passes through a Whatman GFC filter having an effective pore size of *ca.* 1μ.

Method

A. *Capabilities*

Range: 2 to 40 µg-at N/l
Precision at the 20 µg-at N/l level:
  The correct value lies in the range, mean of *n* determination $\pm 0.4/n^{\frac{1}{2}}$ µg-at N/l

B. *Outline of method*

A sample of seawater is oxidized with potassium persulfate under pressure and organic nitrogen is converted to nitrate. The nitrate is analyzed by the Method 1.1.

C. *Special apparatus and equipment*

125-ml screw-cap Teflon bottles (Nalgene Teflon FEP)
Autoclave
Nitrate columns and reagents as in Method 1.1 (Note c).

D. *Sampling procedure and storage*

Samples are collected with acid-cleaned water bottles, rinsed with distilled water to assure the absence of bacterial slime on the sides of the bottles. 50 ml quantities of seawater are filtered through a Whatman GFC glass fiber filter (*ca.* 1 μ pore size) and stored in the dark at $-20°C$ if analysis is not carried out immediately.

E. Special reagents

1. Sodium hydroxide (1.5M)

Dissolve 120 g of analytical grade NaOH in 2 l of distilled water. The solution is stable for several months if it is stored in a tightly closed polypropylene bottle.

2. Oxidizing reagent

Dissolve 6.0 g of twice recrystallized $K_2S_2O_8$ in 100 ml of 1.5M NaOH, stirring with a Teflon-coated magnetic stirrer. The solution is stable up to 8 days in a Pyrex or Teflon bottle in the dark.

3. Hydrochloric acid (1.4M)

Dilute 200 ml of concentrated hydrochloric acid to 1.7 l with distilled water. Find by titration the quantity of this acid that is required to reduce the pH to the range of *ca.* 2.6 to 3.2 of a 40 ml solution of distilled water to which 6 ml of oxidizing reagent has been added; this is determined as $\times$ ml (Note a and Section H).

4. Buffer solution

Dissolve 75 g of NH_4Cl in 400 ml of distilled water. Adjust to pH 8.5 by the addition of concentrated NH_4OH and dilute to 500 ml with distilled water. The solution is stable when stored in a tightly stoppered bottle.

F. Experimental procedure

1. Measure 40 ml aliquots of seawater samples into 125-ml capacity screw-cap Teflon bottles and add exactly 6.0 ml of oxodizing reagent to each bottle.

2. Autoclave with the caps screwed on (Note b) for 30 min at 15 lbs pressure and at *ca.* 120°C.

3. Allow to cool, and add $\times$ ml (see Section E,3) of the 1.4M hydrochloric acid reagent from an automatic pipette adjusted to give exactly $\times$ ml; mix to dissolve the precipitate.

4. Transfer the contents to a 125-ml Erlenmeyer flask and add 3.0 ml of

buffer solution to each flask; mix and analyze the samples for nitrate as described in Method 1.1 (Note c).

5. Measure the extinction of the samples and correct them for the reagent blank. Then the quantity of organic nitrogen is given as:

$$\mu\text{g-at organic N/l} = (E \times F) - A$$

where E is the corrected sample extinction, F is the factor as determined below and A is the μg-at/l of nitrate plus nitrite originally present in the seawater sample (Note d).

G. Determination of blank

Measure 40 ml of distilled water into a 125-ml capacity Teflon bottle and carry out steps 1 to 4 in Section F above. The mean of 3 blanks should be subtracted from the unknown samples and the standard. Since some sources of distilled water give high blanks, it may be necessary to use double distilled water in order not to introduce erroneously high blanks (i.e., $>ca.$ 1 μg-at N/l).

H. Calibration

The factor F in Section F, step 5, is the same as the calibration factor for nitrate when allowance is made for the change in volume. Therefore, F in Method 1.1 (a value of approximately 25) is multiplied by the final volume of solution in the Erlenmeyer flask (Section F, step 4) divided by 40. Thus

$$F = F_N \times \frac{(49+x)}{40}$$

where F_N is the nitrate factor, and x is the number of ml of acid added in step 3 (Section F). The value obtained should be approximately 33.

Notes

(a) pH control in this method is critical. The oxidation must be carried out in the pH range 12.6–13.2; the precipitate formed during the oxidation must be redissolved in the pH range 2.6–3.4 and the nitrate reduction must occur in the pH range 8.0–8.4.

(b) Screwing the caps on prior to autoclaving reduces variability but may result in deformation of the Teflon bottles on cooling. This is only temporary, however, and the original shape returns when air is let in.

(c) For the nitrate determination using a total volume of only 40 ml, the small nitrate columns shown in Fig. 1 are most convenient.

(d) The organic nitrogen estimated will also include ammonia and urea unless these are determined separately and subtracted from the total. Complete conversion of all organic nitrogen to nitrate is generally achieved. However, among the most difficult compounds to oxidize is urea and an investigator may wish to check the oxidation conditions using a urea standard. Conversion of the standard urea under conditions described here should be at least 90%.

2.2. Determination of Dissolved Free Amino Acids by Fluorometric Analysis

Introduction

The method followed in this procedure is an adaptation of the techniques outlined by Benson and Hare (*Proc. Nat. Acad. Sci.*, *USA*, **72**: 619, 1975), Josefsson *et al.* (*Anal. Chim. Acta*, **89**: 21, 1977), Dawson and Pritchard (*Mar. Chem.*, **6**: 27, 1978) and Lindroth and Mopper (*Anal. Chem.*, **51**: 1667, 1979). Free amino acids, peptides, and proteins in seawater are present in trace amounts. The desalting and concentration of large volumes of seawater samples have usually been needed in order to analyze for amino acids. The techniques depend mainly on the sensitivity of the determinations using such methods as spectrophotometry and thin layer and paper chromatography. Gas liquid chromatography is more sensitive than the other methods described above, but the procedures are complex and time consuming. The reason that slow progress has been made in this field is due to the difficulty of having a method for routine analysis, in spite of the fact that amino acids are one of the most important organic constituents from a marine ecological point of view.

Recently, two kinds of fluorogenic reagents (i.e., o-phthalaldehyde and fluorescamin) have been used for the detection of α-amino acids; these replace the use of ninhydrin. In the present method, we adopt the o-phthalaldehyde reagent for the analysis of amino acids in seawater; the reagent is more sensitive than fluorescamin and is stable in aqueous buffers.

Dissolved combined and total particulate amino acids are also measured with the modification of hydrolysis prior to the fluorometric determination of amino acids. Fractional determination of individual amino acids is also possible if high pressure liquid chromatography is used.

Method

A. Capabilities

Range: 0.04–10 µmole/l (3.0–751 µg/l as glycine equivalent)
1. Precision at the 0.5 µmole/l level:
The correct value lies in the range, mean of n determinations $\pm 0.11/n^{\frac{1}{2}}$ µmole/l
2. Precision at the 10 µmole/l level:
Mean of n determinations $\pm 0.31/n^{\frac{1}{2}}$ µmole/l

B. Outline of method

o-Phthalaldehyde, in the presence of 2-mercaptoethanol, reacts with primary amines to form highly fluorescent products. The fluorescent products are measured using a 1-cm cell with a fluorometer. The optimal excitation wavelength (λ_{ex}) is 342 nm, and optimal emission wavelength (λ_{em}) is 452 nm.

When the dissolved amino acids in a water sample are less than the detection limits, the concentration of amino acids in the water sample is carried out using an ion exchange procedure. Seawater samples for amino acid analysis are fractionated by filtering as described in Method 2.4, Section B. The combined dissolved amino acids and total particulate amino acids are analyzed by the methods described in Methods 2.3 and 3.4.

C. Special apparatus and equipment (Note a)

20-ml capacity screw-topped test tubes with teflon liner
200-ml capacity reagent bottles
5-ml automatic pipettes with disposable tips
Rotary evaporator

D. Sampling procedure and storage

The procedures for the collection and filtration of seawater samples are the same as for carbohydrate analysis as described in Method 2.4, Section D. If amino acid concentrations are present in less than detectable limits, about 100 ml of filtered samples should be stored in clean polyethylene bottles after addition of a few drops of concentrated HCl and preserved in a freezer at a temperature below $-20°C$.

E. Special reagents

1. o-Phthalaldehyde solution

Dissolve 200 mg of analytical grade o-phthalaldehyde in 20 ml of 95% ethanol.

2. 0.4M borate buffer solution

Dissolve 24.7 g of reagent grade ortho-boric acid in redistilled water and make up to 1 l. Adjust pH to 9.5 ± 0.02 with 1N NaOH.

3. Buffered reagent solution

Add 1 ml of reagent grade 2-mercaptoethanol to 400 ml of borate buffer. Add 20 ml of o-phthalaldehyde solution to 400 ml of borate buffer containing 2-mercaptoethanol. The buffered reagent solution can be used after 1 hr and is stable for about 2 weeks in a refrigerator.

F. Experimental procedure

a. Direct measurement of free amino acids dissolved in seawater

1. Pipette 5 ml of a seawater sample into a 20-ml capacity screw-topped test tube. Add 5 ml of buffered reagent solution to the test tube and mix well.
2. Let stand for 2 min at room temperature and measure the fluorescence of the sample at $\lambda_{ex} = 342$ nm and $\lambda_{em} = 452$ nm (Note b).
3. Correct the measured fluorescent intensity (F_S) by subtracting the blank value (F_B). Calculate the dissolved free amino acid concentration in μmoles/l from the expression:

$$\mu\text{mole/l} = (F_S - F_B) \times F$$

where F_S is the mean fluorescent intensity of triplicate analyses of water samples, F_B is the mean fluorescent intensity of triplicate analyses of blanks, and F is determined as described below.

b. Pretreatment of water sample if the concentration of amino acids is less than the detection limit (Note c)

1. Prepare an ion exchange column (2×20 cm in size) with 20-ml bed volume of buffered cation exchange resin (Dowex 50X8 H form, pH 2–3) for a sample size of 100 ml.
2. Pass the seawater sample through the column at a flow rate of about 5 ml/min and rinse with 2 or 3 bed volumes of distilled water.

3. Elute with 80 ml of reagent grade 2N NH₄OH into a rotary evaporator flask and reduce the volume under vacuum at 40°C.

4. Transfer the concentrate (*ca.* 10 ml) into a 25-ml capacity rotary evaporator flask and evaporate to dryness at 40°C.

5. Pipette 5 ml of 0.02N NaOH into the flask in order to remove ammonia and evaporate to dryness again at 40°C.

6. Dissolve the sample with 5 ml of 0.02N HCl.

7. Measure the fluorescence of the sample according to Section F–a, 1–3.

G. *Determination of blank*

Determination of the blank for direct measurement of the sample (Section F–a) is carried out using redistilled water instead of a seawater sample, employing the procedure described in Section F–a, 1–3.

The determination of the blank for a concentrated sample is carried out by taking a sample of 100 ml of redistilled water through the whole procedure as described in Section F–b, 1–7, of this chapter.

H. *Calibration*

1. *Standard amino acid solution (Note d)*

Dissolve 0.751 g of glycine in redistilled water and make up to 1000 ml. Store in a refrigerator in glass bottles with 1 ml of toluene. The solution is stable for many months.

$$1 \text{ ml} \equiv 10 \ \mu\text{mole}$$

Dilute 10 ml of this solution in redistilled water and make up to 1000 ml. The diluted solution should be prepared fresh every day.

$$1 \text{ ml} \equiv 0.1 \ \mu\text{mole}$$

2. *Procedure*

Prepare the dilute solutions of glycine (i.e., 0.2, 0.4, 0.6, 0.8, 1.0 μmole/l respectively) and redistilled water for blank determination. Pipette 5 ml of the standard solution into a screw-topped test tube. Carry out the determination exactly as described in Section F for each of the three standard samples and blanks.

A regression line is calculated from the mean fluorescent intensity of the diluted standard solutions by the least squares method. A factor, *F*, is the slope of the regression line expressed as μmole/l/fluorescence unit. This slope is independent of salinity (Note e).

Notes

(a) All glassware should be cleaned with weak alkali solution (i.e., 0.1N NaOH) overnight, dipped in tap water, washed with redistilled water, dried in an oven and capped immediately after drying.

(b) With the glycine standard solution, the fluorescent intensity reaches a maximum 80 to 140 sec after mixing of the sample and reagent. Afterward, the intensity diminishes gradually at a rate of 2.5% per min. The periods corresponding to maximum fluorescence vary slightly depending on the species of amino acid.

(c) Some caution is required to reduce the background fluorescent intensities. These include complete aging of resin, use of highly pure reagents, and the avoidance of contamination from fingerprints on containers or from air-borne particles in the laboratory.

(d) Relative fluorescent intensities for 10 μmole/l of amino acids are as follows (values in parentheses are from data of B. Josefsson): Gly, 10.0; Ala, 9.0; Val, 8.5; Ileu, 7.7; Leu, 8.1; Ser, 10.8; Thr, 8.8; Pro, 0.1; Hypro, 0.0; Asp, 8.2; Glu, 9.6; Lys, 1.5; (His, 8.6); (Arg, 10.6); (Tyr, 11.1); (Phe, 10.3); Met, 9.9; Cys/2, 0.2; (Trp, 11.8); Ammonia, 1.8.

(e) This method has no salt effect in the range of salinity from 35 to 0‰.

2.3. Determination of Dissolved Combined Amino Acids by Fluorometric Analysis

Introduction

For the determination of dissolved combined amino acids, filtered seawater is hydrolyzed by hydrochloric acid and subsequently examined by fluorometric measurement as described in Method 2.2.

The procedures of acid hydrolysis, some additional reagents, and estimation of combined amino acids are discussed. The method is taken basically from procedures published by Dawson and Pritchard (*Mar. Chem.*, **6**: 27, 1978).

Method

A. Capabilities

Range, precision and limit of detection are the same as for the method for dissolved free amino acids as described in Method 2.2, Section A.

B. Outline of method

An aliquot of filtered seawater is hydrolyzed with diluted hydrochloric acid for 22 hr at 110°C. The hydrolyzate is evaporated with a rotary evaporator at a temperature below 40°C and is re-dissolved with distilled water prior to fluorometric analysis. The fluorescence of the hydrolyzate is measured according to the procedure for dissolved free amino acids.

Total dissolved amino acids are determined in this method, so that the combined dissolved amino acids are estimated by the difference of the total dissolved and the dissolved free amino acids, which has been determined already by the method described in Method 2.2.

C. Special apparatus and equipment

    20-ml capacity glass ampules
    Controlled temperature bath
    Rotary evaporator

See also Method 2.2, Section C, with respect to other apparatus and equipment

D. Sampling procedure and storage

Sampling and sample storage are the same as described in Method 2.2, Section D.

E. Special reagents

1. Hydrochloric acid

Reagent grade concentrated hydrochloric acid (sp. gr., 1.19) is used in this experiment.

2. 0.02 N sodium hydroxide solution

Dissolve 0.8 g of reagent grade sodium hydroxide in redistilled water and make up the volume to 1 l.

3. See also Method 2.2, Section E, with respect to the other reagents.

F. Experimental procedure

1. Pipette 5 ml of filtered water into a glass ampule and add the concentrated HCl to make the final acidity 6 Normal.
2. Aerate the sample in the tube with highly pure nitrogen gas for 5 min and seal the tube (Note a).
3. Let stand for 22 hr at 110°C in order to hydrolyze the dissolved combined amino acids in the seawater sample.
4. Cool the sealed ampule with tap water and open the ampule after hydrolysis.
5. Transfer the hydrolyzate into a 25-ml capacity rotary evaporator flask and evaporate to dryness at a temperature below 40°C.
6. Pipette 5 ml of 0.02 N NaOH into the flask in order to remove ammonia in the hydrolyzate and evaporate to dryness again at a temperature below 40°C (Notes b, c).
7. Dissolve the hydrolyzate with 5 ml of redistilled water.
8. Measure the fluorescence of the sample solution according to Method 2.2, Section F, 1–3.
9. Correct the measured fluorescent intensity (F_S) by subtracting the blank value (F_B). Calculate the dissolved amino acid concentration as glycine equivalents in μmole/l from the expressions (Note d):

$$\mu\text{mole/l of total dissolved amino acids (TDA)} = (F_S - F_B) \times F$$

and

$$\mu\text{mole/l of dissolved combined amino acids (DCA)} =$$

$$\mu\text{mole/l TDA} - \mu\text{mole/l DFA (dissolved free amino acids).}$$

G. Determination of blank

Determination of the blank for total dissolved amino acids is carried out by taking a sample of 5 ml of redistilled water through the whole procedure as described in Section F, 1–8, in this method (Note e).

H. Calibration

See the text for dissolved free amino acids as described in Method 2.2, Section H (Note f).

Notes

(a) Acid hydrolysis should be carried out in a nitrogen gas atmosphere to protect from oxidative decomposition of amino acids. The sample is aerated through a capillary glass tube with pure nitrogen gas for *ca.* 5 min prior to sealing the ampule.

(b) Some caution is required to reduce the fluorescent intensity of the blank; for example, removal of ammonia derived from reagents and degradation products of organic nitrogen compounds in the sample.

(c) Removal of ammonia in a dried sample is completed by keeping it in a silica gel desiccator under vacuum after the treatment with weak alkali.

(d) In this method, dissolved combined amino acids are expressed on a molar basis of glycine. A conversion to weight basis should be carried out by multiplying by the mean molecular weight of combined amino acids.

(e) Batches of sealed ampules (usually *ca.* 20 ampules per batch) are prepared in a series of experiments, with a blank and at least one standard solution.

(f) The glycine standard should be analyzed both with and without the hydrolysis procedure to check for contamination from reagents.

2.4. Determination of Total Dissolved Monosaccharides

Introduction

In the present method, total dissolved monosaccharides in seawater and fresh water are determined as an equivalent amount of glucose. The method, which was established by Johnson and Sieburth (*Mar. Chem.*, **5**: 1, 1977), is able to determine the total dissolved monosaccharides in a small sample with high sensitivity and reproducibility. Some modifications have been made to maintain the reproducibility and sensitivity in the procedures.

This method is characterized spectrophotometrically by having similar molar extinction coefficients for various kinds of monosaccharides including hexose, pentose, alditol, and sugar acids, although deoxy sugars have half the molar extinction of the monosaccharides described above.

Method

A. Capabilities

    Range: 0.15–50 μmole/l (27–9000 μg/l as hexose)
    1. Precision at the 10 μmole/l carbohydrate level:
      The correct value lies in the range, mean of n determinations $\pm 0.37/n^{\frac{1}{2}}$

μmole/l (or $\pm 66/n^{\frac{1}{2}}$ μg/l as hexose)
2. *Precision at the 0.3 μmole/l carbohydrate level:*
The correct value lies in the range, mean of n determinations $\pm 0.10/n^{\frac{1}{2}}$ μmole/l (or $\pm 18/n^{\frac{1}{2}}$ μg/l as hexose)

B. Outline of the method

Monosaccharides dissolved in water samples, such as hexose and pentose, are reduced to alditol with potassium borohydride. The alditol originally existing in the sample remains unchanged. The total alditol is oxidized with periodate to form two moles of formaldehyde/mole of monosaccharide. The aldehyde is determined spectrophotometrically with MBTH (3-methyl-2-benzothiazolinone hydrazone hydrochloride). Total dissolved and total particulate carbohydrates are also determined after acid hydrolysis. The scheme is shown below, and the analytical procedures are described in Methods 2.5 and 3.3.
Scheme for separation of dissolved and particulate carbohydrates

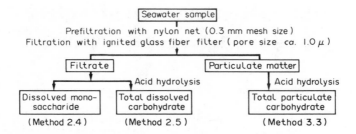

C. Special apparatus and equipment (Note a)

10-ml capacity screw-topped test tubes with teflon-lined caps
Automatic pipettes with disposable tips (e.g., Biopipettes of Schwarz/Mann)
Controlled temperature bath

D. Sampling procedure and storage

Seawater samples collected from various depths are filtered through clean 0.3 mm mesh nylon netting to remove the larger zooplankton. An adequate volume of prefiltered sample is then filtered through an ignited glass fiber filter. The filtrate and particulate matter on the filter (see Method 3.3) are immediately analyzed or stored at a temperature below $-20°C$ for future analyses.

E. Special reagents (Note b)

1. Potassium (or sodium) borohydride solution

Dissolve 100 mg of analytical reagent grade KBH_4 (Note c) in 5 ml of chilled redistilled water (*ca.* 4°C). This solution should be prepared for immediate use.

2. Periodic acid solution

Dissolve 0.57 g of analytical grade periodic acid in 100 ml of redistilled water. This solution should be stored at room temperature in the dark.

3. Sodium arsenite solution

Dissolve 3.25 g of analytical grade sodium arsenite in 100 ml of redistilled water. The solution is stable indefinitely.

4. Ferric chloride solution

Dissolve 5 g of analytical grade ferric chloride in 100 ml of redistilled water, filter through a Whatman GFC filter and store at 5°C.

5. MBTH (3-methyl-2-benzothiazolinone hydrazone hydrochloride) solution

Dissolve 276 mg of analytical grade MBTH (Note c) in 10 ml of 0.1N HCl with warming; filter through a glass fiber filter if precipitates remain. Store in a clean amber bottle at room temperature. This solution should be prepared fresh every day.

6. Acetone

Use analytical grade acetone.

7. 0.7N hydrochloric acid

Dilute concentrated HCl (*ca.* 12N) with redistilled water to make 0.7N HCl.

F. Experimental procedure

1. Sample solutions should be brought to a temperature between about 18° and 20°C. Pipette a 1.0 ml portion (in triplicate) of the water sample into 10-ml

capacity screw-capped test tubes with teflon liners. Pipette 0.05 ml of potassium borohydride solution into each test tube. Allow the solution to stand for 4 hr to complete reduction at 18°C in the dark.

2. Remove the excess borohydride with 0.05 ml of 0.7N HCl. Remove the cap of the test tube and leave for 10 min at room temperature to allow hydrogen gas to evolve. The pH of the solution should be optimal at *ca.* 2.5 for the next step (Note d).

3. Pipette 0.1 ml of periodic acid solution into the sample solution and let stand for 10 min at room temperature in the dark. In this reaction, two hydroxyl groups of alditol are oxidized to aldehyde groups.

4. Pipette 0.1 ml of sodium arsenite solution into the sample solution in order to stop the oxidation reaction. Let stand for at least 10 min in the dark (Note e).

5. Pipette 0.2 ml of 2N HCl into the sample and allow the amber color to disappear.

6. Pipette 0.2 ml of MBTH reagent into the sample solution. Allow the tightly-capped tubes to stand for 3 min in a boiling water bath. Then cool the sample to room temperature with tap water.

7. Pipette 0.2 ml of ferric chloride solution into the sample and let stand for 30 min at room temperature in the dark.

8. Add 1 ml of acetone and mix.

9. As soon as possible, measure the extinction of the solution in a 1-cm cell against 1.8:1 of water:acetone mixture at a wavelength of 635 nm.

10. Correct the measured extinction by subtracting the analytical blank (see Section G). Calculate the total monosaccharide concentration in water samples in μmoles/l from the expression,

$$\mu\text{mole/l} = (E_s - E_b) \times F,$$

where E_s is the mean extinction of triplicate analyses of water samples, E_b is the mean extinction of duplicate analyses of analytical blanks and F is determined as described below.

G. Determination of blank

For analytical blanks, 1.0 ml portions (in duplicate) of the sample are reduced with borohydride and the pH adjusted as described in Sections F–1 and F–2. Subsequently, 0.2 ml of the mixture of sodium arsenite and periodic acid (1:1, V/V) is pipetted into the sample tubes (in place of 0.1 ml of arsenite and 0.1 ml of periodic acid injected separately) followed by Sections F–5 to F–10 (Notes f, g).

H. Calibration

1. Standard monosaccharide solution

Dissolve 0.180 g of glucose in redistilled water and make up to 1000 ml. Store in refrigerator in a glass bottle with 20 mg/l of mercuric chloride. The solution in stable for many months.

$$1 \text{ ml} \equiv 1 \ \mu\text{mole}$$

Dilute 10 ml of this solution with distilled water and make up to 1000 ml. The dilute standard should be made up immediately before analysis.

$$1 \text{ ml} \equiv 10 \times 10^{-3} \ \mu\text{moles} \ (1.8 \ \mu\text{g})$$

2. Procedure

Prepare the dilute solution of glucose (i.e., 10 μmoles/l). Pipette 1 ml of the solution into five test tubes respectively. Carry out the determination exactly as described in Section F for three standard samples, and measure the analytical blanks as described in Section G for two test tubes, respectively. Calculate the factor F from the expression,

$$F = \frac{10}{E_S - E_B}$$

where E_S is the mean extinction of the three standards and E_B is the mean extinction of the two analytical blanks. The value for F is independent of salinity and should be near 24 in μmole/l units (4.3 in mg/l units).

Notes

(a) All glassware used in this experiment should be soaked for a day in weak alkali (0.1N NaOH) and well rinsed with distilled water. Contamination of serine from skin should be avoided as this amino acid can interfere. With the exception of serine, there is negligible interference from naturally occurring organic matter dissolved in water.

(b) Avoid cigarette smoke and formalin during this experiment.

(c) Keep the two reagents, potassium borohydride and MBTH, in a desiccator in a cool and dark place.

(d) At the step of HCl addition prior to oxidation with periodic acid, some modification is introduced with respect to the acidity of the sample solution. This is to assure complete oxidation. Optimal pH of the sample solution should be about 2.5 for the oxidation step.

(e) By separating the reduction and oxidation phase from the colorimetric determination of the resultant formaldehyde, the analyst can process about 20 samples for total monosaccharide every 2 days.

(f) The mixture of sodium arsenite and periodic acid for the analytical blank should be prepared at least 10 min before use.

(g) Three subsamples and two analytical blanks should usually be analyzed on an unknown sample.

2.5. Determination of Total Dissolved Carbohydrates by MBTH Assay

Introduction

For the determination of total dissolved carbohydrates, filtered seawater is hydrolyzed by hydrochloric acid and subsequently examined by MBTH assay as described in Method 2.4. The procedures for hydrolysis, some additional reagents, and estimation of polysaccharides are considered. The method is taken, basically, from a procedure published by Burney and Sieburth (*Mar. Chem.*, **5**: 15, 1977) and Johnson, Burney and Sieburth (*Mar. Chem.*, **10**: 467, 1981).

Method

A. Capabilities

Range: 0.15–50 μmole/l (27–9000 μg/l as hexose)
1. Precision at the 10 μmole/l level:
The correct value lies in the range, mean of n determinations $\pm 1.0/n^{\frac{1}{2}}$ μmole/l

B. Outline of the method

An aliquot of filtered seawater is hydrolyzed with diluted hydrochloric acid. The hydrolyzed sample is neutralized with sodium hydroxide solution, followed by reduction with potassium borohydride. The aliquot of hydrolyzate is pipetted into a test tube and the determination is carried out according to procedures described in Method 2.4. In this method total dissolved carbohydrates are determined spectrophotometrically, so that the dissolved polysaccharides are estimated by the difference between the *total* dissolved monosaccharides and the dissolved monosaccharides which have been measured according to the procedures of Method 2.4.

C. Special apparatus and equipment

20-ml capacity glass ampules
Controlled temperature bath
See also Method 2.4, Section C, with respect to other apparatus.

D. Sampling procedure and storage

Sampling and sample storage are described in Method 2.4, Section D.

E. Special reagents

1. 1N and 2N hydrochloric acid

Dilute analytical grade concentrated HCl (usually 12N) and store in glass bottles.

2. 0.5N sodium hydroxide

Dissolve 20 g of reagent grade NaOH in 1 l of redistilled water and store in a polyethylene bottle.

3. 10% potassium borohydride solution

Dissolve 500 mg of reagent grade KBH_4 in 5 ml of chilled redistilled water. This solution should be prepared for immediate use.

4. See also Method 2.4, Section E.

F. Experimental procedure

1. Pipette 10 ml of filtered water sample and 1 ml of 1N HCl into a cleaned glass ampule (Note a).
2. Seal the ampule with a gas burner.
3. Let stand for 20 hr at 100°C to hydrolyze the polysaccharides dissolved in seawater.
4. Cool and open the ampule after hydrolysis.
5. Transfer the hydrolyzate into a 20-ml capacity test tube.
6. Neutralize the sample with 2 ml of 0.5N NaOH.

7. Pipette in 0.1 ml of chilled 10% potassium borohydride and incubate the mixed solution for 4 hr at 18°C.

8. Pipette 0.2 ml of 2N HCl into the tube and leave for 10 min at room temperature to allow hydrogen gas to evolve.

9. Pipette 1 ml of the sample into a 10-ml capacity screw-topped test tube with teflon liner.

10. See Method 2.4, Section F, 3–10, for the rest of the procedure (Note b).

11. Correct the measured extinction by subtracting the analytical blank. Calculate the total dissolved carbohydrate concentration in the water sample in mg/l from the expression,

$$mg/l = \text{corrected extinction} \times F,$$

where F is a factor determined as described below.

G. Determination of blank

Analytical blanks for total dissolved carbohydrates should be measured according to the procedures described in Method 2.4, Section G, using the hydrolyzate.

H. Calibration

1. Standard polysaccharide solution

Dissolve 0.180 g of good quality soluble starch in redistilled water and make up to 1000 ml. Store in refrigerator in a glass bottle with 1 ml of saturated mercuric chloride solution. The solution is stable for many months.

$$1 \text{ ml} \equiv 180 \ \mu g$$

Dilute 10 ml of this solution with distilled water and make up to 1000 ml. The dilute standard solution should be made immediately before analysis.

$$1 \text{ ml} \equiv 1.8 \ \mu g \text{ (Note c)}$$

2. Procedure

Prepare the dilute solution of soluble starch (i.e., 1.8 mg/l). Pipette 10 ml of the solution and 1 ml of 1N HCl into a 20-ml capacity glass ampule. Allow to hydrolyze and follow subsequent procedures according to Section F, 2–10, in this chapter.

Carry out the measurements for three standards and two analytical blanks.

Calculate the factor F from the expression

$$F = \frac{1.8}{E_s - E_b} \text{ (mg/l)}$$

where E_s is the mean extinction of the three standards and E_b is the mean extinction of the two analytical blanks. The value for F is independent of salinity and should be approximately 4.3 in mg/l units (Note d).

3. Estimation of the concentration of dissolved polysaccharides in seawater

Molar concentration of total dissolved carbohydrates should be determined from the corrected extinction data using a glucose standard (Note e). In order to use the glucose standard for both dissolved monosaccharide (MCHO) and total dissolved carbohydrate (TCHO), extinctions from hydrolyzed samples are multiplied by a factor of 1.19 to correct for dilution of the samples by the hydrolytic and neutralization reagents. Namely, dissolved polysaccharides (PCHO) are estimated from the expression

$$(\mu M\ PCHO) = (\mu M\ TCHO \times 1.19) - (\mu M\ MCHO)$$

or

$$(\mu g/l\ PCHO) = (\mu M\ PCHO \times 162) - (\mu M\ MCHO \times 180)$$

Notes

(a) If care is taken to avoid the loss of sample solution during hydrolysis, a screw-capped test tube with teflon liner can be used instead of a sealed glass ampule.

(b) Removing the excess borohydride by hydrochloric acid should be gently carried out because the contents in the test tube may run over with the evolution of hydrogen gas.

(c) Molarity of a polysaccharide solution refers to that of the glycosidically linked residues. In our experiment, the diluted standard (1.8 mg/l of soluble starch) was approximately equivalent to 10 μmole/l of glucose.

(d) Except Na-alginate and agar, extinctions of 7 kinds of polysaccharides lie between 77.4 and 101.0 (mean 90.8 ± 8.7) compared with 100% as glucose on a molar basis.

(e) Glucose standards should be analyzed both with and without the hydrolysis procedure to check for contaminants of reagents and extra handlings compared with the determination of monosaccharides.

2.6. Determination of Petroleum Hydrocarbons

Introduction

A sensitive analytical method is required for petroleum hydrocarbons because of their low concentration at the ppb level in natural environments. Recently, a simple technique of fluorescence spectroscopy has been developed for the analysis of total petroleum hydrocarbons. The present method is taken basically from Keizer and Gordon (*J. Fish. Res. Bd. Canada*, **30**: 1039, 1973).

Method

A. Capabilities

The limit of detection will depend upon the volume of water extracted and the sensitivity of the fluorometer. Using less than 2 l water samples and a Turner fluorometer at a range setting of 1X and 3X for the determination of petroleum hydrocarbons, the capabilities are as follows:

Range: 5–2000 µg/l
1. *Precision at the 200 µg/l level:*
   The correct value lies in the range, mean of n determinations $\pm 7/n^{\frac{1}{2}}$ µg/l
2. *Precision at the 1500 µg/l level:*
   The correct value lies in the range, mean of n determinations $\pm 110/n^{\frac{1}{2}}$ µg/l
3. *Recovery of hydrocarbons by solvent extraction:*
   Recovery of hydrocarbons is about 87% at the 1000 µg/l level, if the extraction condition is used as described in Section F.

B. Outline of method

Trace amounts of petroleum hydrocarbons in seawater are extracted three times with dimethyl chloride immediately after sampling. The extracts are stored in the dark at a low temperature (5°C) before analysis. The solvent is evaporated to dryness in a rotary evaporator at *ca.* 30°C under reduced pressure. The concentrate is taken up in 10 ml of n-hexane. The fluorescence of the sample is measured at 310 nm excitation and at 374 nm emission wavelength, respectively.

C. *Special apparatus and equipment*

0.2, 1.0 and 2.0-l capacity glass separatory funnels

Rotary evaporator with 200-ml capacity flask

Turner fluorometer with F4T4-BL lamp, Corning 7–54 filter for the excitation light and Corning 7–60 filter for the emission light. If a spectro-fluorometer is available, the corresponding wavelength settings are 310 nm and 374 nm, respectively.

D. *Sampling procedure and storage*

Sample collection should be carried out using a polyethylene bucket or PVC Niskin bottle. Seawater samples are transferred to 2-l polyethylene bottles before analysis (Note a).

E. *Special reagents*

1. Spectrophotometric grade methylene chloride

2. Spectrophotometric grade n-hexane

3. Analytical reagent quality acetone

F. *Experimental procedure*

The analytical scheme for the determination of hydrocarbons is shown in Fig. 2.

1. Pour 1 l of seawater into a 2-l capacity glass separatory funnel and extract three times with 40, 20 and 20 ml of dimethyl chloride, respectively (Note b).

2. Drain the solvent phase into a 100-ml capacity glass bottle and seal tightly with a screw-cap lined with dimethyl chloride-washed aluminum foil.

3. Store the extract in the dark and at a low temperature (5°C) before analysis.

4. Remove any water present in the extract using a 200-ml glass separatory funnel.

5. Evaporate the extract to dryness in a rotary evaporator at *ca*. 30°C under reduced pressure. Dissolve the concentrate in 10 ml of n-hexane.

6. Excite the hydrocarbons in n-hexane at near 310 nm and measure the fluorescence at near 374 nm wavelength. Calculate the concentration of hydrocarbons as follows:

$$\text{hydrocarbon } (\mu g/l) = F_D \times (R_S - R_B) \times v/V$$

where F_D is a factor for each door setting, R_S is the fluorometer reading of the sample, R_B is the fluorometer reading of a solvent blank (see Section G), v is

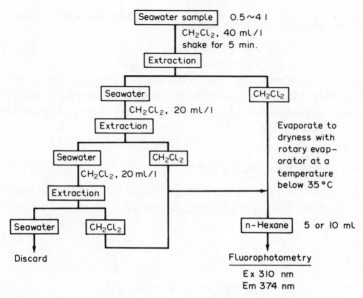

FIGURE 2. Analytical scheme for extraction and measurement of petroleum hydrocarbon.

the volume of hexane in ml, and V is the volume of the original seawater sample in ml (Note c).

G. Determination of blank

Pour 80 ml of dimethyl chloride into a rotary evaporator flask and evaporate to dryness at 30°C. Dissolve the residue in 10 ml of n-hexane and measure the fluorescence exactly as described in Section F. The mean of two blanks should be used as the R_B value.

H. Calibration

Calibrate the fluorescence response with a solution of crude oil dissolved in n-hexane in a concentration range comparable to unknown samples or, if the identity of the oil in seawater is known, a sample of the identical oil should be used for calibration (Note d).

A door factor, F_D, is calculated as follows:

$$F_D = \frac{c}{R}$$

where c is the concentration of crude oil in $\mu g/l$ and R is the fluorometer reading of the concentration (c) of crude oil, with excitation near 310 nm and emission near 374 nm.

Notes

(a) Samplers, containers and all glassware should be cleaned with acid, distilled water, acetone and dimethyl chloride before use.

(b) Availability of dimethyl chloride and the disadvantage of carbon tetrachloride as the extraction solvent for petroleum hydrocarbons have been pointed out by Keizer and Gordon (*loc. cit*) and Ware and Lewis (*J. Chem. Phys.*, **57**: 3546, 1972).

Solvent extraction should be carried out within a few hours after sampling.

(c) This method is not specific for petroleum hydrocarbons but detects any fluorescing compounds that are extractable with dimethyl chloride and are not lost during the evaporation step, such as lipids in living organisms. In the determination of petroleum hydrocarbons in a sample having a high population of organisms, the extract should be cleaned using a silica gel column prior to the measurement of fluorescence in order to remove polar organics which are not found among petroleum hydrocarbons. For the purification, a glass column (1 cm in diameter and 15 cm in length, packed with 20–40 mesh silica gel which was previously washed with n-hexane) is prepared. The concentrated extract dissolved in 3 ml of n-hexane is passed through the column and eluted with n-hexane until exactly 36 ml of total volume is collected. The effluent is evaporated and redissolved with 10 ml of n-hexane. The fluorescence is measured as in Section F, step 6. In this elution, the more polar lipids from the biota are retained on the column when it is eluted with 36 ml of hexane. The above procedure has been proposed by Zsolnay (Second IOC/WMO Workshop on Marine Pollution (Petroleum) Monitoring, Monte Carlo, Monaco, 14–18 June, 1976). Further precise information on hydrocarbon species can be obtained by gas liquid chromatography.

(d) Chrysene (1, 2-benzophenanthrene) can be used as a standard for hydrocarbons instead of crude oil, if the appropriate standard oil is not available. The conversion factor should be estimated previously from the ratio of fluorescence response between chrysene and crude oil. The conversion factor (chrysene/crude oil) is a constant, 7.72 according to Shigehara *et al* (Oceanographical Magazine, *30*, 61, 1979). We compared the fluorescence responses of two kinds of crude oil, provided from Prudhoe Bay and Norman Wells. The standard deviation in both cases was in the range of about 8%. For Norman Wells and Prudhoe Bay oil, under the experimental conditions reported here, F_D was about 22 ng/l per fluorescent unit using × 1 (times one) sensitivity.

SECTION 3

Particulate Material

3.1. Determination of Particulate Organic Carbon

Introduction

The estimation of particulate organic carbon is most accurately carried out by combustion of a filtered sample in oxygen and the measurement of carbon dioxide produced. Various pieces of apparatus are available for this purpose and detailed instructions are given by the manufacturer. If such equipment is not available, a simple method is given here which is based on the use of the spectrophotometer. The method involves the wet oxidation of carbon by acid dichromate and is based on the procedure described by Johnson (*J. Biol. Chem.*, **181**: 707, 1949), adapted for spectrophotometry. The method may give higher results (*ca.* 10 to 20%) than the determination of carbon by measurement of carbon dioxide, due to the oxidation of highly reduced compounds such as lipids. The "oxidizable carbon" as measured by the following technique can be interpreted, in fact, as a measure of the energy stored in a sample of particulate material.

Method

A. Capabilities

Range: 10 to 4000 µgC/l for samples filtered from 10 to 0.5 l
Precision at the 800 µgC/l level:
The correct value lies in the range, mean of *n* determinations $\pm 120/n^{\frac{1}{2}}$ µg carbon

B. Outline of method

The particulate matter is filtered onto a glass fiber filter. Carbon is determined by "wet ashing" with dichromate and concentrated sulfuric acid. The decrease in extinction of the yellow dichromate solution is taken as a measure of the oxidizable carbon.

C. Special apparatus and equipment

4.5 cm Whatman GFC glass filter papers and standard Millipore filtration equipment. The filters must be freed of oxidizable material by placing them on aluminum foil in a muffle furnace at 450–500°C for 24 h (Note a).

50-ml stoppered graduated cylinders

30-ml Pyrex beakers with cover glasses (All glassware to be cleaned in acid dichromate and well rinsed prior to use.)

Block heater or sand bath at 100–110°C

D. Sampling procedure and storage

Samples should be filtered as soon as the seawater is collected and the filters analyzed immediately or stored in a desiccator.

E. Special reagents

1. Sulfuric acid-dichromate oxidant

Dissolve 4.84 g of potassium dichromate, $K_2Cr_2O_7$, in 20 ml of distilled water. Add this solution *a little at a time* to about 500 ml of concentrated (sp. gr. 1.82) analytical quality sulfuric acid in a 1000-ml volumetric flask. Cool the mixture to room temperature and make to volume with more concentrated sulfuric acid. Store in a glass-stoppered bottle protected from dust; the solution is stable indefinitely.

2. Phosphoric acid

Analytical reagent grade (70%) phosphoric acid

3. Sodium sulfate solution

Dissolve 45 g of anhydrous sodium sulfate, Na_2SO_4, in 1000 ml of distilled water.

F. Experimental procedure

1. Place a 4.5 cm glass filter in a Millipore filter holder and attach a controlled vacuum source which cannot exceed about 1/3 atm. After filtration of a suitable sample of seawater (usually 0.5 to 2 l), apply full suction to the filter. Release the suction after 1 min, add 2 ml of sodium sulfate reagent, and reapply the suction immediately; repeat once more with 2 ml of sodium sulfate and remove the filter under suction (Note b).

2. Place the filter in a 30-ml beaker and press into the bottom. Add 1.0 ml of phosphoric acid and 1.0 ml of distilled water. Mix and place in a block heater at 100–110°C for 30 min. Cover with a watch glass during this period.

3. Add suitable volumes of sulfuric acid-dichromate oxidant and distilled water as shown in the following table, depending on the anticipated carbon content:

Anticipated carbon (μg)	Oxidant (ml)	Water (ml)	Final volume (ml)	Cuvette length (cm)
up to 300	2.00	0.8	100	10
300–700	4.00	1.6	50	2.5
700–2000	10.00	4.0	50	1

4. Mix by swirling and place a cover glass over each beaker. Heat for 60 min at 100–110°C.

5. Cool the mixture and transfer the solution and glass fiber filter to a suitably-sized graduated cylinder as indicated in 3 above. Rinse the sides of the beaker with distilled water and make the graduated cylinder up to volume with distilled water. Stopper and mix by inverting; allow to stand at room temperature to cool and to allow the filter to settle in the bottom of the cylinder (Note c).

6. Measure the extinction of a blank solution *against the sample* (Note d) at 440 nm using a cuvette path length as indicated in 3 above.

7. Correct the resulting extinction for the absorbance of trivalent chromium by the expression:

$$E = 1.1 E_f$$

where E_f is the extinction found by difference in 6 above. Calculate the particulate carbon in μgC/l from the expression:

$$\mu\text{gC/liter} = \frac{E \times F \times v}{V}$$

where V is the volume of seawater filtered in liters, v is the volume of oxidant used in step 3 (above), and F is the factor as described in Section H, below.

G. Blank determination

Blank determinations should be carried out with each analysis using filter papers and the quantity of oxidant employed for the unknown samples. Steps 1 to 5 should be carried out in Section F. The blank extinction measured against distilled water should be between 1 and 1.1. The blank should then be used as in step 6, Section F above.

H. Calibration

1. Standard glucose solution

Dissolve 7.50 g of pure glucose and a few crystals of mercuric chloride, $HgCl_2$, in distilled water and make up to a volume of 100 ml. The solution is stable for many months in the refrigerator but should be discarded if any turbidity develops.

Dilute 10.0 ml of the concentrated solution to 1 l with distilled water. Then:

$$1.00 \text{ ml} \equiv 300 \text{ } \mu g \text{ of carbon}$$

2. Procedure

Place one glass fiber paper with 1 ml of phosphoric acid in a 30-ml beaker. Heat for 30 min at 100–110°C. Add 10 ml of oxidant and 4 ml of dilute glucose solution to each beaker. Continue the method as in Section F, steps 4 to 7. Calculate the factor F as:

$$F = \frac{120}{E_S}$$

where E_S is the average of three standard extinctions corrected for the trivalent chromium absorption at 440 nm. The value of F should be approximately 275 (Note e).

Notes

(a) The temperature should not exceed 500°C or the filtering characteristics will be altered.

(b) The washing removes chloride trapped in the filter from seawater. With *rapid* washing there is no significant disruption of cellular material. The further treatment of the filter with phosphoric acid volatilizes remaining chloride to an acceptable level (i.e., less than 0.1 mg chloride will not interfere with the reaction).

(c) If the glass fibers do not settle out readily (especially for the 10-cm cuvette readings), a portion of the mixture can be centrifuged or filtered through another glass fiber filter.

(d) The blank solution, with a higher extinction than the sample, is placed in the spectrophotometer cell normally used for the sample and the sample in the cell normally used for the blank. In this manner, the *difference* in extinction is measured.

(e) The method can also be standardized by titration using 0.05N ferrous ammonium sulfate and a suitable redox indicator.

3.2. Determination of Total Particulate Carbohydrate by Anthrone Reagent

Introduction

In this method, total particulate carbohydrate in a seawater sample is determined by anthrone reagent. This determination is a measurement of the blue-green chromophores coming from furfural derivatives produced as degradation products of carbohydrates when combined with polyaromatic compounds, such as anthrone, N-ethylcarbazol, orcinol, phenol and probably tryptophan. The choice of anthrone for the determination of carbohydrates in particulate organic material was made on the basis of its simplicity, which allows for the analysis of many samples. However, the method is less accurate than the MBTH assay since not as many sugars react to give the same amount of color as glucose, which is a distinct advantage of the MBTH method. The method is basically taken from Hewitt (*Nature*, **182**: 246, 1958); Antia and Lee (*Fish. Res. Bd. Can.*, Ms. Rep. Ser., **168**: 117 pp., 1958) and Strickland and Parsons (*Bull. Fish. Res. Bd. Canada*, **167**, 2nd ed., 231, 1972).

Method

A. Capabilities

Range: 2–4000 µg/l (using between 10 and 0.5 l samples)
Precision at the 500 µg level:
The correct value lies in the range, mean of n determinations $\pm 54/n^{\frac{1}{2}}$ µg glucose

B. Outline of method

Particulate samples, which were filtered through an ignited glass fiber filter with suspended magnesium carbonates, are resuspended with a constant volume of redistilled water. An aliquot of the resuspension is warmed in a boiling water bath with anthrone sulfuric acid reagent. After cooling, the extinction of the resulting blue-green color is measured spectrophotometrically at 625 nm.

C. Special apparatus and equipment (Note a)

30-ml capacity glass test tubes with glass stoppers, *ca.* 15 cm by 1.6 cm in diameter
15-ml graduated stoppered centrifuge tubes

D. Sampling procedure and storage

See Method 2.4, Section D.

E. Special reagents

1. Anthrone reagent

Dissolve 0.20 g of anthrone (9,10-dihydro-9-ketoanthracene) in 100 ml of concentrated analytical reagent quality sulfuric acid (sp. gr. 1.82) and add 8 ml of ethyl alcohol and 30 ml of distilled water. Allow to stand for 4 hr at room temperature or overnight in a refrigerator before use; shade from light at all times (Note b).

2. Sulfuric acid

Use analytical grade sulfuric acid which has a low reagent blank for determinations.

3. Magnesium carbonate suspension

Add 1 g of magnesium carbonate powder [i.e., $(MgCO_3)_4 \cdot Mg(OH)_2 \cdot nH_2O$, 10 μ particle size] to 100 ml of distilled water in a 250-ml Pyrex Erlenmeyer flask fitted with a ground-glass stopper.

F. Experimental procedure

1. Collect particulate matter using a glass fiber filter previously treated with magnesium carbonate powder to prevent particles adhering to the surface of the membrane.
2. Add 10 ml of distilled water to the centrifuge tube containing the particulate matter and $MgCO_3$ (Note h).
3. Pipette 1.0 ml of homogeneous suspension (agitated vigorously with a tube buzzer) into a test tube.
4. Add 10 ml of anthrone reagent to the test tube (Note c).
5. Place the tube in a boiling water bath for 6 min (Note d).

6. Cool the mixture rapidly by placing the tube immediately in ice water for 2 min, then remove.

7. Place tube in a water bath at *ca.* 20°C and leave for 5 min.

8. Measure the extinction in a spectrophotometer, against water, using a 1-cm glass cell at 625 nm.

9. Correct the resulting extinction by a reagent blank (see Section G).

10. Calculate the carbohydrate from the expression:

$$\mu g \text{ glucose/l} = \frac{\text{corrected extinction} \times F}{V}$$

where V is the volume (in liters) of seawater used and F is a factor determined as described in Section H (Notes e, f).

G. Determination of blank

1. Reagent blank

With each batch of samples, a blank determination should be undertaken by carrying out the method exactly as described in Section F on a blank of magnesium carbonate. The blank extinction should not be exceeded by about 0.1.

2. Pigment blank

The correction for a pigment blank is not necessary in the determination of particulate matter in open seawater samples. In samples collected from highly productive areas with large amounts of living phytoplankton, the suspended samples are prepared for the pigment blank analysis.

The procedure is the same as the analysis for carbohydrates described in Section F except for using sulfuric acid instead of the anthrone sulfuric acid solution. The extinction is measured against distilled water and the resulting corrected pigment extinction is subtracted from the sample extinction values.

H. Calibration

1. Standard glucose solution

Dissolve 1.00 g of pure glucose in distilled water and make the volume to 100 ml with distilled water. The solution is stable for a few weeks if stored in a refrigerator at 1–5°C.

$$1 \text{ ml} \equiv 10 \text{ mg glucose}$$

Dilute 10 ml of stock solution to 1000 ml with distilled water. The dilute standard solution should be prepared immediately before use.

$$1 \text{ ml} \equiv 100 \ \mu g$$

2. Procedure

Pipette 1 ml of dilute glucose standard (in triplicate) and distilled water (in duplicate) and add the corresponding amount of 2 ml of magnesium carbonate which is loaded on a glass fiber filter into five test tubes, respectively. Add 10 ml of anthrone reagent to all tubes. Carry out the determination exactly as described in Section F, 4–9, for three standard samples, and measure the reagent blanks as described in Section G for two tubes, respectively. Calculate the factor, F, from the expression (Note g):

$$F = \frac{1000}{E_S - E_B}$$

where E_S is the mean of the three standard extinction values and E_B is the average of the two blank determinations. The factor F should be approximately 2000 and should be determined at least once with each new batch of anthrone reagent.

Notes

(a) All glassware must undergo a preliminary cleaning in weak alkali for at least overnight.

(b) Care should be taken to avoid contamination of the concentrated sulfuric acid from contact with plastic caps, cap liners, cellulose, cellulose wraps, paper strips, straw, etc. before and after use.

(c) Batches of tubes (usually 10–16 tubes per batch) are prepared in a series of experiments, with a blank and at least one sugar standard.

(d) A reaction period for 6 min is the optimal period for the most sensitive aldohexose.

(e) It should be noticed that the percentage absorbance from different sugars at 625 nm relative to glucose indicates the following order of sensitivity: glucose, fructose (100) > sorbose (75) > methyl pentose (70) > galactose (60) > mannose (50) ≫ uronic acids, pentose (3–4).

(f) The 470 and 550 nm ratios expressed as percent of the extinction at 625 nm can be roughly averaged for sugar types as follows: hexose, 30, 45; methylpentose, 15, 60; pentose, 220, 130; hexuronic acids, 140, 200, respectively.

(g) 1.0 ml of the homogeneous suspension is taken for analysis since this allows for other analyses to be performed for particulate material on the same sample. However, if only carbohydrates are being analyzed, then the particulate residue can be resuspended in 1 ml of distilled water and the anthrone reagent added directly to the resuspended material. Note that the factor, F, is approximated as 2000 to allow for one-tenth of the original material on the filter. If all the filtered material is used, the factor F is approximated as 200.

3.3. Determination of Total Particulate Carbohydrate by MBTH Assay

Introduction

In this method, the total particulate carbohydrate collected on a filter (Whatman GFC) is hydrolyzed by hydrochloric acid and subsequently determined by MBTH assay as described in Method 2.4. This is the application of dissolved carbohydrate as described by Burney and Sieburth (*Mar. Chem.*, **5**: 15, 1977).

Method

A. Capabilities

Range, precision and limit of detection are the same as for the method for total dissolved monosaccharides described in Method 2.4, Section A.

B. Outline of method

A particulate sample, prepared according to Method 2.4, Section B, is hydrolyzed in a sealed glass tube with 10 ml of 0.1 N HCl for 20 hr at 100°C in a constant temperature bath. After hydrolysis, the sample is filtered through an ignited glass fiber filter. The filtrate is neutralized with sodium hydroxide solution followed by reduction with potassium borohydride. A 1-ml aliquot of hydrolyzate is pipetted into a 10-ml screw-topped test tube and the determination made according to MBTH assay (Method 2.4).

C. Special apparatus and equipment

20-ml capacity glass ampules
Controlled temperatue bath
Filtering apparatus with fritted glass of 25 mm diameter
See also Method 2.4, Section C.

D. Sampling procedure and storage

Seawater samples of 0.1 to 1.0 l should be filtered through ignited glass fiber filters (i.e., Whatman GFC). The method of sample storage is mentioned in Method 2.4, Section D.

E. Special reagents

Special reagents used in this experiment are the same as those in Methods 2.4 and 2.5, Section E except for 0.1N HCl.

F. Experimental procedure

1. Put a particulate sample, together with the glass fiber filter, and 10 ml of 0.1N HCl into an ampule.
2. Seal the top of the ampule with a gas burner.
3. Let stand for 20 hr at 100°C.
4. Cool and open the sealed tube.
5. Filter through an ignited glass fiber filter.
6. Transfer the filtrate into a test tube.
7. Neutralize the filtrate with 2 ml of 0.5N NaOH.
8. See Method 2.5, Section F (7–10) and Method 2.4, Section F (3–10) for remaining procedures.
9. Correct the measured extinction by subtracting the analytical blank. Calculate the total particulate carbohydrate concentration in the water samples in mg/l from the expression

$$\text{mg/l} = \frac{(E_s - E_b) \times F}{V}$$

where E_s is the mean extinction of triplicate analyses of water samples, E_b is the mean extinction of duplicate analyses of analytical blanks, V is the volume in liters of seawater used, and F is a factor determined as described in Method 2.5, Section H.

G. Determination of blank

Determination of the analytical blank for total particulate carbohydrates should be made according to the procedures described in Method 2.4, Section G.

H. Calibration

See the calibration for total dissolved carbohydrates described in Method 2.5, Section H.

3.4. Determination of Total Particulate Amino Acids by Fluorometric Analysis

Introduction

Total particulate amino acids in seawater are determined by the modified method of dissolved combined amino acids as described in Method 2.3. Here, the procedures of acid hydrolysis and some additional reagents are given.

Method

A. Capabilities

Range, precision and limit of detection are the same as in the method for dissolved free amino acids described in Method 2.2, Section A.

B. Outline of method

A particulate sample collected on a glass fiber filter is hydrolyzed in a sealed glass tube with 6N HCl for 22 hr at 110°C. The hydrolyzate is filtered with a glass fiber filter and evaporated to dryness at *ca.* 40°C under reduced pressure using a rotary evaporator. The dried sample is redissolved with an appropriate volume of redistilled water and the fluorescence intensity measured.

C. Special apparatus and equipment

Filtering apparatus with fritted glass of 25 mm diameter
See also Method 2.3, Section C.

D. Sampling procedure and storage

An adequate volume of pre-filtered samples (0.1 to 1.0 l) is filtered through an ignited glass fiber filter (i.e., 25-mm size Whatman GFC). The method of sample storage is mentioned in Method 2.4, Section D.

E. Special reagents

1. 6N hydrochloric acid solution

Dilute analytical grade concentrated hydrochloric acid with redistilled water.

2. See also Method 2.3, Section E.

F. Experimental procedure

1. Place the filter on which the particulate material has been collected in an ampule containing 5 ml of 6N HCl.
2. Aerate with nitrogen gas for 5 min to replace dissolved oxygen.
3. Seal the ampule with a gas burner.
4. Let stand for 22 hr at 110°C to hydrolyze the particulate proteinaceous substances.
5. Cool the sealed ampule with tap water and open after hydrolysis.
6. Filter the hydrolyzate through a glass fiber filter and wash the ampule contents with redistilled water. Filtrate is collected in a 25-ml capacity rotary evaporator flask (Note a).
7. Evaporate to dryness at a temperature below 40°C.
8. Pipette 5 ml of 0.02N NaOH into the flask in order to neutralize the sample and remove ammonia in the hydrolyzate; evaporate to dryness again at a temperature below 40°C.
9. Dissolve the hydrolyzate with 5 ml of redistilled water.
10. Measure the fluorescent intensity of the sample solution according to Method 2.2, Section F.
11. Correct the measured fluorescent intensity (F_S) by substituting the blank value (F_B). Calculate the particulate amino acid concentration as glycine equivalents in μmole/l from the expression (Note b):

$$\mu\text{mole/l of total particulate amino acids} = (F_S - F_B) \times F \times \frac{v}{V}$$

where v is the volume of redistilled water and V is the volume of seawater, both in ml.

G. Determination of blank

Determination of the blank for total particulate amino acids is carried out by taking a sample ignited glass fiber filter through the whole procedure as described in Section F, 1–11, in this method.

H. Calibration

See the text on dissolved free amino acids as described in Method 2.2, Section H.

Notes

(a) When the sample solution is colored markedly after hydrolysis, amino acids in the sample are purified using a cation exchange resin as described in Method 2.2, Section Fb, 1–7.

(b) The other comments written in Method 2.3 are valid also in the analysis of total particulate amino acids.

3.5. Determination of Particulate Lipid

Introduction

The method described here is an application of the technique outlined by Pande, Khan and Venkitasubramanian (*Anal. Biochem.*, **6**: 415, 1963). The principle involved is the oxidation of lipid with acid dichromate. Using spectrophotometry, it is possible to determine microgram amounts of lipid in the particulate matter of seawater.

The method is not specific for lipid, as other organic matter also may be oxidized by acid dichromate as described in Method 3.1. Therefore, a pretreatment of the particulate sample is carried out by solvent extraction according to Bligh and Dyer's method (*Can. J. Biochem. Physiol.*, **37**: 911, 1959), which was modified by Jeffries (*Limnol. Oceanogr.*, **14**: 41, 1969).

Method

A. Capabilities

   Range: 2–140 µg/l
   1. Precision at the 10 µg/l level:
   The correct value lies in the range, mean of n determinations $\pm 4.0/n^{\frac{1}{2}}$ µgl
   2. Precision at the 100 µg/l level:
   The correct value lies in the range, mean of n determinations $\pm 5.8/n^{\frac{1}{2}}$ µg/l

B. Outline of method

The principle of this method depends upon the oxidation of lipid by acid dichromate. The oxidation reaction is followed by a decrease in the dichromate color (micromethod) or the liberation of iodine from iodide which is added after the oxidation reaction (ultra-micromethod). Therefore, in both the micromethod and ultra-micromethod, the extinction of the reaction medium has an inverse relationship based on the decrease of the dichromate color or a release of iodine, respectively.

C. Special apparatus and equipment

   50-ml stoppered Erlenmeyer flask
   50-ml pear-shaped separatory funnel
   25-ml capacity glass homogenizer with teflon rod
   Motor for homogenizer
   Rotary evaporator with 25-ml volume flask
   Constant temperature bath

D. Sampling procedure and storage

Seawater samples are collected by Niskin bottles which are well washed by dilute hydrochloric acid, distilled water and acetone. One-half to 10 l of seawater are filtered through an ignited glass fiber filter (e.g., Whatman GFC) immediately after sampling. The extraction of lipid should be commenced as soon as possible. If longer storage periods are necessary, the samples should be stored in a freezer at *ca.* $-20°C$ immediately after filtering.

E. Special reagents

Ultra-micromethod for 2–15 μg of lipid:

1. 0.034% (W/V) potassium dichromate in concentrated sulfuric acid (freshly prepared)

Dissolve 0.170 g of analytical grade potassium dichromate in 10 ml of redistilled water. Make the solution to 500 ml with analytical quality concentrated sulfuric acid (sp. gr. 1.82).

2. Cadmium iodide starch reagent

Dissolve 1.4 g of cadmium iodide (CdI_2) in 60–70 ml of water and boil the solution for 15–20 min. Add a clear solution of starch obtained by boiling 0.33 g soluble starch in about 75 ml of water for 5 min. Mix the solution and boil for 5 more min. Filter through a glass fiber filter and make up to 500 ml with distilled water. Store the solution in a glass bottle in the dark. The solution is stable for many months.

Micromethod for 15–140 μg of lipid:

3. 0.15% potassium dichromate in concentrated sulfuric acid

Dissolve 0.75 g of analytical grade potassium dichromate in 10 ml of water with heating; cool the solution. Make the solution to 500 ml with analytical grade concentrated sulfuric acid (sp. gr. 1.82).

Organic solvent for lipid extraction:

4. Mixture of chloroform, methyl alcohol and water

Mix 200 ml of analytical grade chloroform, 400 ml of analytical grade methyl alcohol and 160 ml of distilled water.

5. Chloroform

Use analytical grade chloroform

F. Experimental procedure

I. Extraction of lipid from particulate materials in seawater

1. Filter particulate material onto a glass fiber filter and place in a glass homogenizer. Add 8 ml of mixed organic solvent ($CHCl_3$: MeOH: H_2O = 1:2:0.8 V/V) to the tube. Homogenize the sample, immersing the tube in cold water for at least 1 min. Add 2 ml of chloroform and 2 ml of distilled water. Repeat with blank filter as a control.

2. Filter through an ignited glass fiber filter (see Method 3.1). Pour the filtrate into a 50-ml capacity separatory funnel. Collect the lower layer (chloroform layer) in a 25-ml capacity rotary evaporator flask and evaporate to dryness *in vacuo*.

3. Dissolve the dried lipid sample in a small volume of chloroform (< 2 ml). Transfer the solution to a clean glass stoppered flask. Place the tube in a vacuum desiccator and reduce the pressure gradually by adjusting the

stopcock to avoid loss of lipid fraction until the solvent is completely removed (Note a).

II. *Oxidation of lipid by acid dichromate solution*

(a) Ultra-micromethod (Note b):

1. Add 1.0 ml of 0.034% acid dichromate solution to the sample and lipid-free (blank) flask, respectively.

2. Place the flasks in a boiling water bath for 15 min (Note c) and then cool the solution in running tap water.

3. Add 9 ml of distilled water and mix well. Pipette 0.5 ml of the solution and add 4.5 ml CdI_2-starch reagent. Mix the solution thoroughly and allow to stand at room temperature for exactly 20 min; then dilute the solution with 5 ml of distilled water (Note d).

4. Measure the extinction of a blank solution *against the sample* (Note e) at 575 nm using a cuvette of 1-cm path length.

(b) Micromethod:

1. Add 2.0 ml of 0.15% acid dichromate solution to the sample flask and lipid-free (blank) flask, respectively.

2. Place the flasks in a boiling water bath for 15 min (Note c) and then cool the solution in running tap water.

3. Add 4.5 ml of distilled water. Mix the solution thoroughly.

4. Measure the extinction of a blank solution *against the sample* (Note e) at 440 nm using a cuvette of 1-cm path length.

5. Calculate the particulate lipid in μg/l from the expression:

$$\mu\text{g lipid (as stearic acid)}/\text{l} = \frac{E \times F}{V}$$

where E is the extinction of a blank solution against the sample, F is the factor as described in Section H, below, and V is the volume of seawater filtered in liters.

G. Blank determination

Blank determinations should be carried out with each analysis using filter papers and the quantity of oxidant employed for the unknown samples. Steps a(1)–a(3) and b(1)–b(3) should be carried out in Section F. The blank should then be used as in step a(4) and b(4), Section F, above.

H. Calibration

1. Standard stearic acid solution

Dissolve 0.500 g of pure stearic acid in 500 ml of 95% analytical grade ethyl alcohol.

$$1 \text{ ml} \equiv 1000 \ \mu g \text{ stearic acid}$$

Store in a well stoppered glass bottle. The standard solution is stable indefinitely in a dark and cool place. Dilute 50 ml of this solution to 500 ml with 95% analytical grade ethyl alcohol.

$$1 \text{ ml} \equiv 100 \ \mu g \text{ stearic acid}$$

Store in a well stoppered glass bottle. The dilute standard solution is prepared immediately prior to use.

2. Procedure

Add 1.0 ml of stearic acid standard to a clean flask. Evaporate the solution carefully to dryness with a rotary evaporator at $< 50°C$ and allow to stand in a vacuum desiccator overnight. Following the ultra-micromethod or the micromethod as described in Section F above, determine three extinctions from the standard solution, using three separate 1.0 ml aliquots, against the reagent blank as described in Section G.

Calculate factor F from the expression:

$$F = \frac{C}{E}$$

where C is the concentration of stearic acid in $\mu g/l$ and E is the average extinction of the blank against a standard sample solution at 575 nm in the ultra-micromethod and 440 nm in the micromethod, respectively. The value of F should be approximately 35 in the ultra-micromethod and 280 in the micromethod, respectively.

Notes

(a) Prior to the oxidation of lipid with acid dichromate, complete removal of the solvent is required. Precaution is necessary to avoid lipid loss caused by too rapid volatilization.

(b) The spectrophotometric determination of iodine liberated by the action of acid dichromate offers some disadvantage over the micromethod due to the instability of the iodine released. However, the former method is more sensitive. Optimal color development is obtained with the use of 0.21N acid strength.

(c) The oxidation occurs in the first 8 min after heating. A heating period of 15 min is adequate, although subsequent heating in the micromethod is not critical.

(d) The color in the micromethod remains stable for a few hours. In the ultra-micromethod, however, the measurement should be carried out 20 min after the addition of CdI_2-starch reagent.

(e) The blank solution, with a higher extinction than the sample, is placed in the spectrophotometer cell normally used for the sample and the sample in the cell normally used for the blank. In this manner, the difference in extinction is measured.

3.6. Determination of Adenosine Triphosphate (ATP) in Particulate Material

Introduction

The amount of ATP in particulate material is a valuable indication of the biomass of live microorganisms, including bacteria, phytoplankton and microzooplankton. The method is particularly useful as a microtechnique to be applied on samples taken from below the euphotic zone and in sediment samples, where it is very difficult to see the living material amongst so much detritus. The method described here is taken from Holm-Hansen and Booth (*Limnol. Oceanogr.*, **11**: 510, 1966) with some modifications suggested by Bulleid (*Limnol. Oceanogr.*, **23**: 174, 1978).

Method

A. Capabilities

Range: 1–4000 ng/l (depending on the volume of seawater filtered)
Precision at the 50 ng/l level:
The correct value lies in the range, mean of n determinations $\pm 5/n^{\frac{1}{2}}$ ng/l

B. Outline of method

Particulate material is filtered from seawater and the residue extracted with a hot buffer. When an aliquot of the extract is added to a mixture of luciferin and luciferase (an extract from firefly lanterns), light is emitted proportionally to the amount of ATP extracted; the emission of light is measured using a sensitive photomultiplier.

C. Special apparatus and equipment

Filtering apparatus to remove microscopic particles from seawater onto 25-mm diameter filters
20-ml beakers with cover glasses
Micropipettes to deliver 0.200 ml
Automatic pipette with 5 ml range
Heater block for 100°C extraction
Graduated test tubes (or centrifuge tubes) to contain 10 ml
Suitable photometer and integrator timer (e.g., Chem-Glow Photometer model J4-7441 and Integrator Timer model J4-74622 manufactured by the American Instrument Company and used in this technique – or, equipment manufactured by SAI Technology, 4060 Sorrento Valley Blvd., San Diego, California)

D. Sampling procedure and storage

Samples should be collected in sterile Niskin water samplers or at least in clean, alcohol-rinsed Van Dorn bottles. The water samples should be processed as soon as possible, but the frozen extracts (see Section F) can be stored for several months and analysis completed on shore.

E. Special reagents

1. Tris buffer

Dissolve 7.5 g trishydroxymethylaminomethane in 3000 ml of distilled water. Adjust to pH of 7.7 to 7.8 by the dropwise addition of 20% HCl. The solution should be dispensed into 250-ml flasks, autoclaved at 15 psi for 15 min and the contents of one or more flasks used for each analysis as required (Note a).

2. Firefly extract

The extract can be purchased (e.g., from Worthington Biochemical Company) in a desiccated form in vials. It is reconstituted with 5 ml of distilled water, giving a suspension of pH 7.4 in 0.05M potassium arsenate buffer and 0.02M $MgSO_4$, all contained as part of the vial contents. (In some cases, it may be desirable to dilute the contents to 10 ml with potassium arsenate and

magnesium sulfate to give more enzyme preparation.) The enzyme preparation should be kept in the dark at *ca.* 10°C *for 2–3 h prior to use.*

F. Experimental procedure

1. Filter between 0.25 and 1 l of sample through a 25 mm Millipore GS
filter (0.22 μ pore size) and suck the filter dry but do not wash (Note b).
2. Put the filter on the bottom of a 20-ml beaker containing 3 ml of boiling
Tris buffer (Note c).
3. After extracting for 3 min (Note d), decant the beaker into a clean dry
test tube (or graduated centrifuge tube). Rinse the beaker and filter with two
washings of *ca.* 0.5 ml of distilled water and make the final volume up to 4 ml
with a few drops of distilled water; cover the tube with parafilm. Mix and cool
the contents in an ice bath (Note e).
4. When the analysis is performed, extract and enzyme preparation should
be at room temperature. Pipette 0.2 ml of extract and 0.2 ml of enzyme into a
photometer cuvette, mix well by inverting over parafilm and place in the
photometer cell holder. Exactly 15 sec after first mixing the contents, measure
the number of intensity units on the integrator timer for exactly 10 sec (or for a
specific time interval).
5. Correct the value obtained for the blank and estimate the ATP as

$$\text{ng ATP/l} = \frac{C_{10} \times F}{R \times V}$$

where C_{10} is the count obtained in 10 sec, F is the standardization factor, R is
the recovery factor, and V is the volume of seawater filtered in liters.

G. Determination of blank

Pipette 0.2 ml of Tris buffer and 0.2 ml of enzyme extract into a cuvette and
record the count exactly 15 sec after mixing, integrated for 10 sec. Subtract this
value from the sample and standard values.

H. Calibration

1. Standard ATP solution

Dissolve 12.3 mg of disodium triphosphate, $C_{10}H_{16}N_5O_{13}P_3Na_2 \cdot 4H_2O$, in a
liter of distilled water. Store frozen in a glass bottle; the solution is stable for
many months.
Dilute 1.0 ml to 100 ml with Tris buffer.

$$0.2 \text{ ml} \equiv 20 \text{ ng ATP}$$

Carry out the procedure as described in Section F, 4 to 5 above, using 0.2 ml of standard ATP and 0.2 ml of enzyme solution. Correct the standard count for the blank and calculate F from the average of three standards as:

$$F = \frac{400}{C_s}$$

where C_s is the average standard count. For the instrument used in this experiment (Chem-Glow Photometer), the value of F is approximately 0.4. However, small differences in F will be encountered with each enzyme preparation and therefore the method has to be restandardized often.

2. Calculation of R (recovery)

Gessey and Costerton (Native Aquatic Bacteria: Enumeration, Activity and Ecology, ASTM STP 695, *Amer. Soc. Test. Mat.*, 1979, p. 117) have shown that recovery of ATP from spiked samples following the extraction procedure was less than 100%. Therefore, in order to allow for this loss, 20 ng of added ATP standard should be carried through the extraction procedure (Section F, 1–5) using a sample of known ATP content which has been analyzed separately. Then the factor R is given as:

$$R = \frac{(\text{spiked ATP} + \text{sample ATP}) - (\text{sample ATP})}{(\text{spiked ATP})}$$

This factor should not be less than about 0.75 (Note f).

Notes

(a) Some materials, particularly if the procedure is applied to sediment samples, have been found difficult to extract for ATP analysis. Bulleid (*loc. cit.*) recommends the following extraction buffer to be used instead of Tris:

McIlvaine buffer: 0.04M Na_2HPO_4, adjusted to pH 7.7 with 0.02M citric acid.

In addition, sulfuric acid-EDTA extraction of sediment material has been recommended as another alternative by Karl and LaRock (*J. Fish. Res. Bd. Canada*, **32**: 599, 1975).

(b) 0.25 l of seawater is generally adequate in coastal areas.

(c) ATP may decompose rapidly following filtration; therefore, extraction must be carried out immediately.

(d) Extraction is not immediate but samples should not be extracted for longer than 3 min, followed by washing.

(e) At this stage, the extracts can be stored for several months in the dark at $-20°C$.

(f) The estimation of R and F have been separated in this procedure in order to detect possible sources of error. For the experienced analyst, F may be determined on a standard which is taken through the whole procedure as a spiked sample.

3.7. Electronic Counting and Sizing of Particles

Introduction

The Coulter Counter® (or equivalent equipment) was originally designed for counting blood cells. Its use in marine science has been widespread both for counting and sizing plankton and for sediment samples. There are several models of Coulter Counters; the Model TA II is the most versatile for work on plankton and sediments but it is also expensive relative to the earlier Model B and the more recent Model Z_{B1}. The following procedure applies to all three models, but particularly to the Z_{B1} which has manual controls. The method is taken from *A Practical Manual on the Use of the Coulter Counter in Marine Science* by Sheldon and Parsons (1966 publ. Coulter Electronics Sales Co., Canada) and from Sheldon and Parsons (*J. Fish. Res. Bd. Canada*, **24**: 909, 1967). It is assumed in this description that the equipment has been set up and made operational by a company representative.

Method

A. Capabilities

1. Precision:

The precision of counts depends on the number of particles counted, regardless of the size of the aperture. The following serves as a guide to the precision of different counts, ± 2 standard deviations.

$$2000 \pm 65 \ (3\%)$$
$$200 \pm 12 \ (6\%)$$
$$70 \pm 11 \ (16\%)$$
$$30 \pm 8 \ (27\%)$$

2. Range:

The range of counts depends on the size of the aperture used. Below a certain value for each aperture tube, the number of particles counted will cause a decrease in precision as indicated above. However, at some higher value, coincidence can occur as indicated by Fig. 3 and discussed under section C, Calibration. Thus for each tube there is an optimum counting range, the lower limit of which will be decided by the actual number of particles counted and what is considered by the investigator to be acceptable confidence limits, as indicated above under Precision. The following values give an indication of the upper limit of particle counts for 4 tubes based on 10% coincidence

1000 μ diameter aperture 100 particles/ml
500 μ diameter aperture 500 particles/ml
200 μ diameter aperture 10,000 particles/ml
100 μ diameter aperture 100,000 particles/ml

B. Outline of method

The apparatus specified in the following procedure is a Coulter Counter Model Z_{B1}. A detailed description of the electronics of the instrument is given

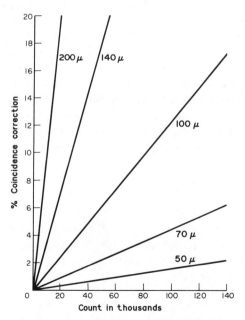

FIGURE 3. Coincidence correction chart for different sized apertures (taken from *Instruction and Service Manual for the Model 3 Coulter Counter®*)

in a manual supplied by the manufacturer. The instrument consists of an electronic cabinet and a seawater sampling stand. The former contains a digital counter display, an oscilloscope screen, upper and lower threshold controls and aperture current and amplification switches. The sampling stand consists of an aperture tube (of different sizes), a mercury manometer, control stopcock, two electrodes, a vacuum pump and a 35x microscope for viewing the aperture for possible blockage by debris.

A vacuum created by the pump is used to draw a sample of electrolyte through the aperture. Particles contained in the electrolyte displace an amount of fluid proportional to their size and number, and this is sensed in the aperture as a change in the electric current. The latter is maintained by having one electrode outside the aperture tube and one inside. The number of particles in a known volume of electrolyte can be determined by using the manometer as a metering device. This is accomplished by allowing the mercury to flow back through the manometer after applying a vacuum. In so doing the mercury passes start and stop contacts which activate the counter to enumerate particles in a known volume. An alternative procedure, especially recommended for large apertures and shipborne operations, is to employ a time switch in place of the mercury column and to count particles for a known time. The quantity of electrolyte passed during a set time-interval can then be determined from the flow rate for a particular aperture.

The upper and lower thresholds act as electronic gates, above and below which no particles are counted. Their operation can be thought of as being analogous to a counting grid on a microscope. The aperture current and amplification switches control the size of particle seen on the oscilloscope – again by analogy, their operation can be thought of as the magnification of the different objective lenses on a microscope. At each change of setting, the aperture current switch doubles (or halves) the current, and the amplification switch doubles (or halves) the instrument response. For the purpose of this text, the combined effect of the aperture current and amplification settings has been expressed as a multiple and called the "sensitivity".

The oscilloscope provides a visual picture of the particles as they pass through the aperture and the effects of the upper and lower thresholds appear as darkened bands on the screen.

The instrument used by us was stable over a temperature range of 5–25°C and a salinity range of 8–40‰ but these values should be checked for a particular calibration. At higher or lower temperatures and salinities, the instrument should be calibrated specifically for the medium employed.

When seawater is drawn through an aperture, an electrical field exists in the immediate proximity of the aperture. If a particle passes through this field, an amount of electrolyte is displaced and the conductivity of the field is altered. Thus both the number of particles and their size will alter the conductivity. By manipulating the threshold controls and sensitivity as described below, the investigator can differentiate between these two properties and record each, separately.

C. Calibration

Calibration of different sized apertures may be carried out with any uniformly sized particles of known volume having a diameter between 2 and 40% of the aperture being calibrated. In practice it has been found convenient to use pollen grains or plastic spheres. The former can be purchased from Hollister-Stier Laboratories, P.O. Box 14197, Dallas, Texas, USA, in sizes ranging from 16 to 90 μ (e.g., paper mulberry pollen, 16 μ; ragweed pollen, 19.5 μ; pecan pollen, 45 μ; corn pollen, 90 μ). The plastic spheres of very small sizes (e.g., 1–10 μ) can be obtained from the Dow Chemical Co., Midland, Michigan, USA. For the largest apertures it is necessary to use particles such as crab eggs (400 μ).

Particles should be suspended in membrane-filtered seawater by shaking, and left for 24 hr to separate and become uniformly hydrated. The diameter of the material employed for calibration should be checked under a microscope and the procedure carried out as follows:

1. Set the lower threshold at 5 and the upper threshold at 100. Adjust amplification and aperture current settings so that the particles appear one-half to two-thirds of the way up the oscilloscope screen.

2. Count the number of particles in a known volume and express the count as the number of particles per milliliter.

3. Adjust the number of particles in the seawater sample so that there are less than the number causing a 5% coincidence (see Fig. 3). For aperture sizes not shown in Fig. 3, coincidence may be approximated from the expression:

$$\frac{(D_1)^3}{(D_2)^3} \times C_2$$

where D_1 is the diameter of the new aperture and D_2 is the diameter of the aperture for which the coincidence (C_2) is known.

4. Set the lower threshold to 5 and the upper to 10. Count the number of particles twice. Repeat at 5 threshold intervals, counting between 10 and 15, 15 and 20 . . . etc., up to 85 and 90.

5. Plot the mean of each count as shown in Fig. 4 and determine the mode of the first peak. Subsequent peaks may occur in this plot but these are due to two or more particles being stuck together.

6. Note amplification and aperture current settings. Divide the volume of the particle used for calibration by the threshold value for the mode.

Example: The example shown in Fig. 4 is for the calibration of a 30-μ aperture using plastic spheres of 22.5 μ^3 volume. The calibration factor obtained at a sensitivity of 0.5 is 0.64 μ^3. If the sensitivity is changed, e.g. to 4, each threshold setting is then equivalent to $(4/0.5) \times 0.64 = 5.1 \mu^3$.

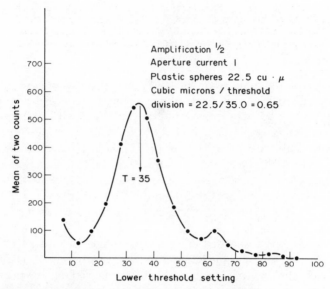

FIGURE 4. Size distribution of plastic spheres used for the calibration of a 30-μ aperture (from Strickland and Parsons, *Bull. Fish. Res. Bd. Canada*, **167**, 1972).

D. Preparation of sample

1. Suspended materials

For naturally occurring suspended materials, it will generally not be necessary to concentrate or dilute seawater samples over the size range 3–100 μ. For particles larger than 100 μ in diameter, the material may be concentrated on a sieve. Particles in the size range 100–500 μ should be concentrated with a 50-μ sieve; larger particles are best collected with a 300-μ mesh diameter net towed through the water; and the largest particles (i.e., greater than 2000 μ) are best removed with a sieve and discarded. The degree of concentration will depend on the environment but, for particles greater than 300 μ in diameter, a thousandfold concentration will generally be necessary. Netted samples are resuspended in a known volume of membrane-filtered seawater. Resuspension is readily accomplished by ordinary mixing but, for particles greater than 300 μ in diameter, vigorous stirring is necessary. In addition, an enlarged seawater reservoir is required to allow for the greater volume of sample drawn through the larger apertures.

For phytoplankton cultures, it may be necessary to dilute the sample in order to avoid coincidence corrections. From a preliminary count of the sample, and with reference to Table 1, it may be determined if the count exceeds an acceptable level of coincidence. Generally, counts which have a

TABLE 1. A grade scale based on 1-μ particle diameter in which particle volume varies by a factor of 2 in successive grades (from Sheldon and Parsons, *A Practical Manual on the Use of the Coulter Counter® in Marine Science*, Coulter Electronic Sales Co., Canada, 1967).

Diameter (μ)	Log$_{10}$D	Volume (μ^3)	Volume (μ^3)	Log$_{10}$D	Diameter (μ)
1.00	0.00	0.52	0.73	0.05	1.12
1.26	0.10	1.04	1.47	0.15	1.41
1.58	0.20	2.08	2.94	0.25	1.78
2.00	0.30	4.19	5.92	0.35	2.24
2.52	0.40	8.38	11.8	0.45	2.82
3.18	0.50	16.8	23.8	0.55	3.57
4.00	0.60	33.5	47.4	0.65	4.49
5.04	0.70	67.0	94.7	0.75	5.66
6.34	0.80	134	189	0.85	7.12
8.00	0.90	268	379	0.95	8.98
10.1	1.00	536	758	1.05	11.3
12.7	1.10	1.07×10^3	1.52×10^3	1.15	14.3
16.0	1.20	2.15×10^3	3.03×10^3	1.25	18.0
20.2	1.30	4.29×10^3	6.07×10^3	1.35	22.6
25.4	1.40	8.58×10^3	12.1×10^3	1.45	28.5
32.0	1.50	17.2×10^3	24.3×10^3	1.55	35.9
40.3	1.60	34.3×10^3	48.6×10^3	1.66	45.3
50.8	1.71	68.7×10^3	97.1×10^3	1.76	57.0
64.0	1.81	137×10^3	194×10^3	1.86	71.9
80.6	1.91	275×10^3	388×10^3	1.96	90.5
102	2.01	549×10^3	777×10^3	2.06	114
128	2.11	1.10×10^6	1.56×10^6	2.16	144
161	2.21	2.20×10^6	3.11×10^6	2.26	181
203	2.31	4.40×10^6	6.22×10^6	2.36	228
256	2.41	8.79×10^6	12.4×10^6	2.46	287
322	2.51	17.6×10^6	24.9×10^6	2.56	361
404	2.61	35.1×10^6	49.6×10^6	2.66	452
512	2.71	70.3×10^6	99.4×10^6	2.76	573
646	2.81	141×10^6	199×10^6	2.86	724
812	2.91	281×10^6	397×10^6	2.96	909
1020	3.01	562×10^6			

coincidence correction of greater than 5% should be diluted. Dilution to a known volume should be made with membrane-filtered (0.22 μ) seawater.

When natural seawater samples are being counted with small apertures, it is necessary to remove much of the larger material in order to avoid repeated blocking of the aperture and, in some cases, the possibility of larger particles masking the presence of smaller particles. For this purpose, sieves should be employed, which will in no way affect the distribution and number of the smaller particles being measured. It is recommended that sieve sizes no smaller than from 1 to 1.5 times the aperture size be employed for this purpose.

2. Sedimented material

Recently deposited inorganic particulate material may be readily sized with a Coulter Counter®. The procedure followed in preparing the samples depends to some extent on the approximate grain size of the particles. In general, however, ignition in a furnace removes organic material from larger particles (100 μ); the addition of hydrogen peroxide with warming on a hotplate to aid digestion cleans smaller particles; and a small amount of dilute hydrochloric acid removes carbonates. In the latter case, excess acid should be removed by washing before sizing the material in the counter.

Particles are readily resuspended by agitation for 5 min in an ultrasonic bath. Samples should be resuspended in a saline solution, or in glycerol-saline if the particles are larger than 100 μ in diameter. The following chart provides a guide to the treatment of any particular sediment sample.

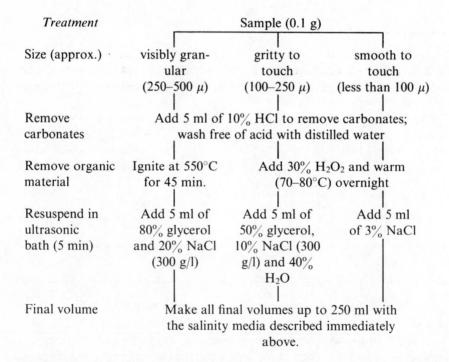

Treatment	Sample (0.1 g)		
Size (approx.)	visibly granular (250–500 μ)	gritty to touch (100–250 μ)	smooth to touch (less than 100 μ)
Remove carbonates	Add 5 ml of 10% HCl to remove carbonates; wash free of acid with distilled water		
Remove organic material	Ignite at 550°C for 45 min.	Add 30% H_2O_2 and warm (70–80°C) overnight	
Resuspend in ultrasonic bath (5 min)	Add 5 ml of 80% glycerol and 20% NaCl (300 g/l)	Add 5 ml of 50% glycerol, 10% NaCl (300 g/l) and 40% H_2O	Add 5 ml of 3% NaCl
Final volume	Make all final volumes up to 250 ml with the salinity media described immediately above.		

In combination with the procedure described above, the use of screens also may be necessary in making preliminary separations of sediments. In this case, a combination of one or more of the procedures shown in the above flow chart may be useful.

E. Procedure for determining particle size spectrum

The choice of an appropriate particle size scale on which to express the distribution of particulate material depends on the range of particle sizes encountered. For particles with a small size range, the distribution of sizes may be expressed on an arithmetic scale of particle diameter (along the abscissa), and as either numbers or volume of particles (along the ordinate). This type of distribution is best suited for unicellular algal cultures or occasionally for natural populations of a monomictic bloom. However, for the great range of particle sizes encountered in most natural populations of suspended materials, as well as among sedimented particles, it is necessary to express particle size on a logarithmic scale of particle diameter (along the abscissa), and the quantity of material *as the volume not number* (along the ordinate). A discussion of the use of this type of size spectrum has been given by Sheldon and Parsons (*J. Fish. Res. Bd. Canada*, **24**: 909, 1967), and a suitable grade scale to employ for the Coulter Counter® is give in Table 1.

1. Procedure – arithmetic distributions

(a) Select an aperture tube such that the particles being examined have a diameter between 2 and 40% of the aperture size.
(b) Follow the procedure described in Section C.
(c) Plot the mean of two counts as the number of particles per unit volume of medium versus the diameter of the particle in each size category. The latter scale is obtained by multiplying the threshold scale (such as is shown in Table 1) by the calibration factor for the particular sensitivity employed, and determining the diameter of the volume obtained, assuming the particles to be spherical (see Note below).

2. Procedure – logarithmic distributions

(a) Select an aperture tube such that the particles being examined have a diameter between 2 and 40% of the aperture size (Note a).
(b) Set the sensitivity at the minimum setting, i.e. for counting the smallest particles (see Note b), and the lower threshold at the lowest level used for making counts. (In the example given in Note b this threshold value would be 46.3.) Set the upper threshold at 100.
(c) Count the particles in a suitable volume of medium and express the count as the number per milliliter. Adjust the volume of the medium if the count exceeds 5% coincidence (Fig. 3).

Note: In both procedures for arithmetic and logarithmic distributions, it is recommended that the particle diameter be used as a measure of size. In doing this, it is recognized that the property actually measured is volume and that the diameter given is that of a sphere having the same volume as the particle.

(d) Set the upper and lower thresholds at the required values for obtaining the number of particles in the size categories shown in Table 1 (Note b) and make two counts at each sensitivity setting until the maximum sensitivity is reached. (e) Express the mean of two counts as the number of particles in a known volume of medium (e.g., 1 ml) and multiply the count by the geometric mean volume for each size interval (the geometric mean volume is given in column 4 of Table 1). The results may then be plotted as the volume of material found in the size categories shown in the sixth column of Table 1. These results give amount as a logarithmic progression of particle diameters (Note c). An example of a logarithmic particle size distribution is given in Fig. 5.

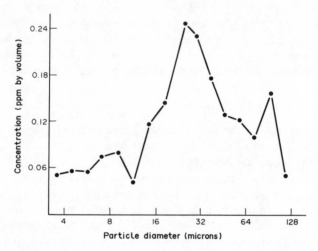

FIGURE 5. An example of a logarithmic size distribution of suspended particulate material (from Strickland and Parsons, *Bull. Fish. Res. Bd. Canada*, **167**, 1972).

Notes

(a) In this procedure it is often necessary to use two or more apertures in order to cover a wide range of particle sizes. The stepwise description given here is for the use of one aperture, and the use of a second larger or smaller aperture will follow the same procedure. Since the grade scale given in Table 1 is continuous, the portion of the scale covered by one aperture may be joined up with an adjacent portion of the scale which is covered by a second aperture.

(b) The required threshold settings have to be determined only once for each tube, providing there are no subsequent changes made to the electronics of the counter. From information obtained under Section C (Calibration), the particle volume corresponding to one threshold setting at a known sensitivity will have already been obtained. It is now necessary to determine the threshold settings (upper and lower) which will give any two adjacent volumes in the left hand column of Table 1. The following example serves to show how this is done:

Suppose that with a 200 μ aperture, one threshold setting corresponds to 1.48×10^3 μ^3 at a sensitivity of 64 (i.e., aperture current 1, amplification 64). Then, at this sensitivity, particles in the size range 68.7×10^3 to 137×10^3 μ^3 (see Table 1) could be measured if the upper and lower thresholds were placed at:

$$T = \frac{137 \times 10^3}{1.48 \times 10^3} = 92.5 \text{ (upper)}$$

and

$$T = \frac{68.7 \times 10^3}{1.48 \times 10^3} = 46.3 \text{ (lower)}$$

Since the movement of the amplification or aperture current settings corresponds to a $2 \times$ volume scale, it is clear that the logarithmic scale of particle diameter given in Table 1 can now be followed by leaving the upper threshold setting at 92.5 and the lower threshold at 46.3, and changing the sensitivity by a factor of 2 each time a count is taken. (Note: Either amplification or aperture current controls may be employed to change the sensitivity.) In this case, the minimum sensitivity will be found from the diameter in Table 1 corresponding to 2% of the aperture size, and the maximum sensitivity will be found from the diameter in Table 1 corresponding to 40% of the aperture size.

(c) By designing a suitable data form, the calculations associated with the conversion of particle numbers in a certain volume, to particle volume in a standard volume of medium can readily be carried out with a computer.

Addendum: The measurement of phytoplankton growth rates

Phytoplankton growth rates of natural populations can be measured using the Coulter Counter® and logarithmic distributions. A discussion of one technique is given by Cushing and Nicholson (*Nature*, **212**: 310, 1966) and an example of this technique will be found in the *Handbook of Phytological Methods*, Ed. J.R. Stein, *Publ.* Cambridge University Press, 1973, 349–358. A discussion of this technique is given by Sheldon (*Limnol. Oceanogr.* **24**: 760, 1979).

3.8. Sinking Rate of Phytoplankton and Other Particulates

Introduction

In many studies, there is a need to know the sinking rate of phytoplankton and other particulates. There are a number of methods which can be employed, but one of the simplest and most accurate is described here as given by Bienfang (*Can. J. Fish. Aquat. Sci.*, **38**: 1289, 1981). A review of other methods will be found in this reference.

The biomass index used to determine sinking can be varied (e.g., particulate carbon, nitrogen, number of particles, chlorophyll, etc.) depending upon the sinking rate property required. In the following description, the method has been applied to chlorophyll as a measure of the sinking rate of phytoplankton.

Method

A. Capabilities

Range: 0.5 to 50 m/day (Note a)

Precision at the 1.8 m/day level:

The correct value lies in the range, mean of n determinations $\pm 0.13/n^{\frac{1}{2}}$ m/day

B. Outline of the method

The method employs a settling column initially containing a uniformly mixed population of cells. Sinking rate is calculated from a change in the vertical distribution of cells after a given time. The ascent rate of positively buoyant materials can also be determined by the same method.

C. Special apparatus and equipment

The design of the column is shown in Fig. 6 (Note b). The column may be constructed of Plexiglas. For a total volume of *ca.* 350 ml, the positioning of the two lateral drainage ports should allow for *ca.* 25 ml to be retained in each section (V_s and V_b, see Section E for symbols). The cap on the top is designed to sit snugly to improve column stability, and it should have a small bore hole to allow air to enter when the column is drained. The column (or columns)

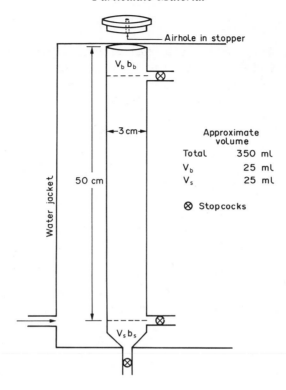

FIGURE 6. Settling column for particulate material. Redrawn from Beinfang (*Can. J. Fish. Aquat. Sci.*, **38**: 1289, 1981). See text for details of lettering.

must be contained in a water jacket which is constructed of 6.4 mm Plexiglas sheets. Water is circulated from a thermostatically controlled bath. A manageable unit of four columns in a single water jacket can be constructed for replication and simultaneous measurements on different samples.

In addition to the settling columns, a rapid method is usually needed to measure the biomass of material. The chlorophyll fluorometric method (in relative fluorescent units, since absolute concentration is not required) or the Coulter Counter method are suggested (see Methods 4.3 and 3.7). A long handled rubber policeman is required to resuspend settled material.

D. Sampling procedure and storage

Samples must be tested for sinking rates immediately after collection in order to avoid any subsequent physiological change in this property.

E. Experimental procedure

1. Fill the water column with the seawater sample containing the phytoplankton to be tested. Set aside a 100 ml aliquot of the same sample which should be placed in a similar environment to the sample being tested on the column. Stopper the column and note the time.

2. Determine the biomass concentration in the water which has been placed in the column (b_{oo}). After time (t), determine the biomass in the 100 ml aliquot which must remain mixed (b_{ot}) (see Note c).

3. After time t (Note d), draw off the upper volume (V_b) by releasing the upper stopcock (flow rate should be < 50 ml/min). The volume, V_b, is then measured precisely and the biomass concentration, b_b, is determined as chlorophyll.

4. The volume, of water in the column is then slowly decanted through the lower lateral opening until only the volume V_s remains. The conical sides of this portion of the tube are lightly scraped to resuspend material on the sides (a long handled rubber policeman serves for this purpose).

5. The volume in the conical part of the tube is then decanted through the bottom stopcock; the volume is measured accurately (V_s) and the concentration of chlorophyll is determined (b_s).

6. Calculate the total biomass (B_t) and the amount settled (B_s) as:

$$B_t = \tfrac{1}{2}(b_{oo} + b_{ot})V_t$$

and

$$B_s = V_s b_s - \tfrac{1}{2}(b_{oo} + b_{ot})V_s$$

Then the sinking rate is given as:

$$S = \frac{B_s}{B_t}\frac{1}{t} \text{ m/day}$$

where l is the height of the water column in meters as indicated on Fig. 6, t is the settling time during the experiment in days, and V_t is the total volume of the column.

Similarly, for particles which tend to float, the ascent rate, A, is given by:

$$A = \frac{B_b}{B_t}\frac{1}{t}$$

where

$$B_b = V_b b_b - \left(1 - \frac{tS}{1_b}\right)\left[\frac{V_b(b_{oo} + b_{ot})}{2}\right],$$

1_b is the length of the V_b region, and B_t and S are as defined above.

Example calculation (Note f):

Using an *in vivo* measure of chlorophyll fluorescence (Note e), the following values were obtained:

$$b_{oo} = 50 \text{ units} \qquad \text{time} = 0.083 \text{ day}$$
$$b_{ot} = 53 \text{ units} \qquad 1 \quad = 0.5 \text{ m}$$
$$b_s \;= 110 \text{ units} \qquad V_t \quad = 355 \text{ ml}$$
$$\qquad\qquad\qquad\qquad\quad V_s \quad = 26 \text{ ml}$$

Then:

$$B_t = 18282.5$$
$$B_s = 2860 - 1339 = 1521$$

and

$$S \;= \frac{1521}{18282.5} \times \frac{0.5}{0.083} = 0.5 \text{ m/day}$$

Notes

(a) The range given here is for a column of 0.5 m in which the shortest sinking time accurately observable is considered to be 0.5 hr and the longest, 24 hr. Obviously by either making longer or shorter columns, or by improving the precision of very short or very long observations, the range can be expanded.

(b) For oligotrophic waters, it may be necessary to construct larger columns having approximate dimensions of 9 cm diameter and a volume of *ca.* 4 l.

(c) The measurement of b_{ot} is to determine if there has been any change in b_{oo} during the time period, t, due to cell multiplication, etc. The two values are later averaged.

(d) The actual time interval to get a measurable difference in b_b and b_s will be found experimentally but may range from *ca.* 0.5 (for large cells) to 24 hr (for very small cells).

(e) The use of relative *in vivo* fluorescence saves considerable time, but it can only be used on uniform crops of phytoplankton, or cultures, since *in vivo* fluorescent yield per cell may change with species. However, the concentration of particulates may also be measured as $\mu g/l$, number/l or in any other concentration units.

(f) In this experiment, columns were placed in the dark. The location of columns in the light or dark may affect the results obtained.

SECTION 4

Plant Pigments

4.1. Determination of Chlorophylls and Total Carotenoids: Spectrophotometric Method

Introduction

The most useful chemical method for determining the total quantity of phytoplankton in seawater is to estimate the amount of chlorophyll (usually as chlorophyll *a*). This value can then be related either to the total biomass of plant material using a factor, or it can be employed in determining the assimilation index in conjunction with photosynthetic rate measurements (Method 5). The first spectrophotometric method for algal pigments in seawater was described by Richards with Thompson (*J. Mar. Res.*, **11**: 156, 1952); some improvements to this method were suggested by Parsons and Strickland (*J. Mar. Res.*, **21**: 155, 1963) and new spectrophotometric equations were given by Jeffrey and Humphrey (*Biochem. Physiol. Pflanzen.*, **167**: 191, 1975). The technique described here has been obtained from the above references, using the spectrophotometric equations for chlorophylls given in the last reference.

Method

A. Capabilities

The limit of detection of plant pigments in seawater cannot be given since it depends upon the total quantity of seawater filtered. If it is not practical to filter more than about 10 l of seawater, then the lower limit is about 0.02 mg/m^3 chlorophyll *a*; if, on the other hand, clogging of filters due to a large amount of detrital material occurs after 1 l has been filtered, then the lower limit of chlorophyll *a* determination will be about 0.2 mg/m^3.

Chlorophyll a *precision at the 5 μg level:* the correct value from *n* determinations lies in the range, $\pm 0.21/n^{\frac{1}{2}}$ μg chlorophyll *a*. The precision for the determination of other chlorophylls and carotenoids is discussed under Notes e and f.

B. Outline of the method

A known volume of seawater is filtered onto a synthetic filter (e.g., Millipore

101

AA) or onto a glass fiber filter; pigments are extracted from the filter in 90% acetone and their concentration is estimated spectrophotometrically.

C. Special apparatus and equipment

Filter equipment designed to hold 47-mm diameter synthetic (e.g., Millipore) membrane or glass fiber filters.

One 300-ml wash bottle for 90% acetone.

Stoppered centrifuge tubes, 15-ml capacity if membrane filters are used or 25-ml capacity fritted glass vacuum filter if glass fiber filters are used.

Spectrophotometer and 10-cm light path spectrophotometer cells to hold 10 ml or less of acetone extract.

Centrifuge for 15-ml tubes.

D. Sampling procedure and storage

Between 0.5 and 10 l of seawater are filtered through a membrane or glass fiber filter (pore size *ca.* 0.5μ). As the seawater is being filtered, a few drops of a suspension of magnesium carbonate in seawater are added to prevent acidity on the filter. The filter is drawn dry, removed, and can be folded and stored in a desiccator at $-20°C$ for at least 30 days if analysis can not proceed immediately. Filters should be folded in half, backed with a piece of ordinary paper and fastened with a paper clip for storage.

E. Special reagents

1. 90% acetone

Pipette 100 ml of distilled water into a 1-l volumetric flask and make up to 1000 ml with analytical grade acetone. The reagent can be conveniently stored in a dark, tightly stoppered bottle and dispensed from a 300-ml wash bottle.

2. Magnesium carbonate

Add *ca.* 1 g of finely powdered $MgCO_3$ to 100 ml of distilled water; shake vigorously before use and dispense a few drops from a 100-ml wash bottle.

F. Experimental procedure

1. Invert a polyethylene bottle containing the seawater sample into the Millipore filtering equipment containing a membrane or fiber glass filter (Note a). Allow the sample to filter under $\frac{1}{2}$ atmospheric pressure vacuum.

2. Add several (3 to 5) drops of $MgCO_3$ solution to the seawater as it is being filtered.

3. Drain the filter thoroughly with the suction and store (Section D) or extract as necessary.

4. Place the filter in a 15-ml centrifuge tube; add 15 ml (Note b) of 90% acetone to volume and shake thoroughly. Allow to stand overnight in a dark place (preferably refrigerated).

5. Centrifuge the contents of each tube at room temperature for 5 to 10 min—the exact time depending on the model of centrifuge and the degree of clarity obtained (optical density at 750 nm should be less than 0.05 in a 10-cm cuvette).

6. Decant the supernatent into a 10-cm path length spectrophotometer cuvette and measure the extinction at the following wavelengths without delay (sample should be at room temperature to avoid misting on the optical cell).

Wavelengths: 750, 664, 647, 630, 510 and 480 nm. (If only chlorophylls are required, then the 510 and 480 nm readings need not be taken.) (See Note c).

7. Correct each extinction for a small turbidity blank by subtracting the 750 nm from the 664, 647 and 630 nm absorptions. (The 510 nm absorbance is corrected by subtracting 2X the 750 nm absorbance and the 480 nm absorbance is corrected by subtracting 3X the 750 nm absorbance.)

8. Calculate the amount of pigment in the original seawater sample using the equations given below:

For chlorophylls,

$$(Ca) \text{ Chlorophyll } a = 11.85 \, E_{664} - 1.54 \, E_{647} - 0.08 \, E_{630}$$
$$(Cb) \text{ Chlorophyll } b = 21.03 \, E_{647} - 5.43 \, E_{664} - 2.66 \, E_{630}$$
$$*(Cc) \text{ Chlorophyll } c = 24.52 \, E_{630} - 1.67 \, E_{664} - 7.60 \, E_{647}$$
$$(*\text{see Note d})$$

where E stands for the absorbance at different wavelengths obtained above (corrected by the 750 nm reading) and Ca, Cb and Cc are the amounts of chlorophyll (in $\mu g/ml$ if a *1-cm* light path cuvette is used); then:

$$\text{mg chlorophyll/m}^3 = \frac{C \times v}{V \times 10} \text{ (Note e)}$$

where v is the volume of acetone in ml (15 ml), V is the volume of seawater in liters and Ca, Cb and Cc are the three chlorophylls which are substituted for C in the above equation, respectively (Note: $\mu g/l \equiv mg/m^3$).

For carotenoids,

$$(C_P) \text{ Plant carotenoids} = 7.6 \, (E_{480} - 1.49 \, E_{510})$$

where E is the absorbance at 480 and 510 nm (corrected for the 750 nm absorbance) and where Cp is substituted for C in the same equation as is used above for chlorophylls (Note f).

Notes

(a) The advantage of membrane filters is that they dissolve in the 90% acetone. However, some phytoplankton are difficult to extract and, in this case, the plankton should be collected on a glass fiber filter. The filter is then placed in a small volume motorized homogenizer and about 8 ml of 90% acetone are added. The filter is homogenized for 30 sec to 1 min and the homogenate is then filtered through a fritted glass filter. The residue is washed with 1 or 2 ml of 90% acetone and the clear filtrate is made up to volume (15 ml) with 90% acetone. The filtrate is then transferred to a spectrophotometric cell and measurements made at the wavelengths described under Experimental Procedure, step 6.

(b) 15 ml of 90% acetone are used here as the standard extraction volume so as to provide sufficient extract for the 10-cm spectrophotometric cell. However, if this cell requires less than 10 ml to fill it, the method can be made more sensitive by adding only 10 ml of acetone.

(c) The method depends on highly accurate settings of the wavelengths on the spectrophotometer and these should be checked against standard hydrogen line emissions.

(d) There are two chlorophylls, C_1 and C_2. In this equation, the chlorophyll *c* represents total chlorophyll *c*.

(e) Lorenzen and Jeffrey (UNESCO Technical Papers in Marine Science, No. 35, 1980) have estimated the errors involved in the use of these equations. For chlorophyll *a*, they found < 1% error; for chlorophyll *b*, < 1% error; for chlorophyll *c*, 24% too low when compared with combinations of pure pigment. Greater errors are introduced if the wavelength of the spectrophotometer is not properly set.

(f) The amount of carotenoids can only be approximated since they are a mixture of several compounds which may have different molar extinction coefficients.

4.2. Spectrophotometric Determination of Phaeo-pigments

Introduction

Chlorophyll degradation products may at times form a significant fraction of the total plant pigment in a seawater sample. The degradation products result from the digestion process of zooplankton which converts chlorophyll into phaeo-pigments (phaeophorbide and phaeophytin) as well as from decomposition processes due to hydrolytic enzymes in the phytoplankton which may convert chlorophyll into chlorophyllide. The absorption coefficient of the latter pigment is the same as the parent chlorophyll and it can not, therefore, be detected by spectrophotometric analysis. The absorption spectrum of the phaeo-pigments is considerably lower in the 665 nm region when compared with the parent chlorophyll. Thus it is possible to determine the amount of phaeo-pigments by measuring the extinction at 665 nm before and after destruction of all chlorophyll in the sample to phaeo-pigments. The method is described for chlorophyll *a* and other chlorophylls can not be assessed separately. Their presence causes a small error in the estimate, as does the presence of phaeo-pigments in the trichromatic equations described in the previous section (Method 4.1). Several methods have been suggested for this analysis (e.g., Moss, *Limnol. Oceanogr.*, **12**: 335, 1967; Lorenzen, *Limnol. Oceanogr.*, **12**, 343, 1967; Marker, *Freshwat. Biol.*, **2**: 361, 1972). The procedure employed here uses equations given in the Lorenzen reference; a comparison of the Lorenzen and Marker methods is given by Lorenzen and Jeffrey, UNESCO Technical Papers in **Marine Science, 35**: 1980.

Method

A. Capabilities

The limit of detection and precision of this method will be similar to the previous method (Method 4.1). However, the accuracy of the method may be considerably less since no corrections are made for other chlorophylls. In ocean waters where chlorophyll *a* and *c* dominate, this is not a serious problem. However, the presence of chlorophyll *b* in lake or riverine systems would be a serious contaminant.

B. Outline of the method

The extinction of an acetone extract of plant pigment is measured before and after treatment with dilute acid. The change following acidification is used as a measure of the quantity of phaeo-pigments in the original sample.

C. Special apparatus and equipment

Same as Method 4.1.

D. Sampling procedure and storage

Same as Method 4.1.

E. Special reagents

1. *Reagents 1 and 2, Method 4.1.*
2. *Hydrochloric acid:* Dilute 10 ml of concentrated hydrochloric acid to 100 ml with distilled water.

F. Experimental procedure

1. Carry out procedures in Method 4.1, steps 1 to 5.
2. Measure the extinction of the extract at 665 and 750 nm. Add 2 drops of dilute hydrochloric acid to the cuvette (Note a), mix (Note b) and remeasure the extinction at 665 and 750 nm.
3. Subtract each 750 nm reading from the corresponding 665 nm extinction and use the following equations to calculate the concentration of chlorophyll *a* and phaeo-pigments in the sample:

$$\text{chlorophyll } a \text{ (mg/m}^3) = \frac{26.7(665_o - 665_a) \times v}{V \times 1}$$

$$\text{phaeo-pigments (mg/m}^3) = \frac{26.7(1.7[665_a] - 665_o) \times v}{V \times 1}$$

where 665_o is the extinction at 665 nm before acidification, 665_a is the extinction at 665 nm after acidification, v is the volume of acetone extract(ml), V is the volume of water filtered (liters) and 1 is the path length of the cuvette (cm).

Notes

(a) If the extraction has been made using a membrane filter, the addition of a small amount of hydrochloric acid may cause an increase in turbidity (measured at 750 nm). This is undesirable if it occurs and it may be better to follow the procedure for using glass fiber filters.

(b) The samples are best mixed by holding a small piece of aluminum foil over the mouth of the cuvette and inverting several times. The destruction of chlorophyll *a* to phaeophytin is pH-dependent and best occurs between pH 2.6 and 2.8 after 3 to 5 min. The choice of the correct acidity avoids breakdown products of phaeophytin and fucoxanthin which can interfere with the determination. Thus the experimenters should check the pH of their extracts following any modification in procedure which they might employ with respect to extracting agent or the addition of an organic base (Moed and Hallegraeff, *Int. Rev. Ges. Hydrobiol.*, **63**: 587, 1978).

Rinse the cuvette thoroughly with 90% acetone after each determination to ensure that no acid is carried over when the next 665 nm reading is taken.

Addendum

Recently, plant pigment analyses have been performed by High Precision Liquid Chromatography (HPLC). While the cost of equipment for HPLC analysis falls outside the scope of this text, it must be noted that HPLC analysis, while more costly and time consuming, is much more accurate than the methods used here.

4.3. Fluorometric Determination
of Chlorophylls

Introduction

The method described here is based on the use of the Turner fluorometer as suggested by Yentsch and Menzel (*Deep-Sea Res.*, **10**: 221, 1963) and subsequently investigated by Holm-Hansen *et al.* (*J. Cons. perm. int. Explor. Mer*, **30**: 3, 1965).

Other literature on the fluorometric technique describes a method (Loftus and Carpenter, *J. Mar. Res.*, **29**: 319, 1971) for the simultaneous determination of chlorophylls *a*, *b* and *c*. The method requires two additional filters for emitted light. From their results, it is apparent that such a trichromatic fluorometric method would be superior to the method described here if (a) a measure of chlorophyll *b* and *c* were required, or (b) the ratio of chlorophyll *b:a* in a sample was greater than 0.2. In the latter case, the presence of chlorophyll *b* could decrease the apparent quantity of chlorophyll *a* and, at the same time, lead to a spuriously high value for phaeo-pigments as determined in Method 4.4. However, in general, the chlorophyll *b:a* ratio in seawater is appreciably below 0.2.

More recently, Boto and Bunt (*Analyt. Chem.*, **50**: 392, 1978) have improved the method of Loftus and Carpenter (*loc. cit.*) by using variable monochromatic excitation instead of narrow band filters.

Method

A. Capabilities

The limit of detection will depend upon the volume of water filtered and the sensitivity of the fluorometer. In general, the method is 5 to 10 times more sensitive than the spectrophotometric method, but it may be less accurate. A precision for replicate measurements at the 0.5 mg/m^3 chlorophyll *a* level should be better than 10%.

B. Outline of method

An acetone extract is made as in Method 4.1 and the extract is measured on a fluorometer.

C. Special apparatus and equipment

Glass fiber filters
Turner fluorometer with F4T4-BL lamp, Wratten 47B or Corning CS.5–60 filter for the excitation light and Corning CS.2–64 filter for the emitted light (Note a).

D. Sampling procedure and storage

See Method 4.1.

E. Special reagents

See Method 4.1.

F. Experimental procedure

1. Extract 0.1 to 2 l of seawater as described in Method 4.1 using a *glass fiber* filter and adding *10 ml* of 90% acetone for the total volume of extract.
2. "Zero" the Turner fluorometer against 90% acetone and measure the fluorescence of the sample.
3. Then:

$$\text{mg chlorophyll } a/\text{m}^3 = F_D \times R \times \frac{v}{V}$$

where R is the fluorometer reading, F_D is a factor for each door setting (see Section G), v is the volume of acetone extract in ml, and V is the volume of seawater in liters (Note b).

G. Calibration

The method should be calibrated against a known concentration of chlorophyll *a* as determined by the spectrophotometric method using an extract of phytoplankton from a vigorously growing culture.

Extract about 50 ml of culture, measure the amount of chlorophyll *a* by the spectrophotometric method, and measure the fluorescence of the same sample using the *least* sensitive door setting. Then the factor (F_D) for that door (3) is:

$$F_D{}^3 = \frac{Ca}{R_3}$$

where Ca is the equivalent concentration of chlorophyll *a* in the acetone extract expressed in μg/ml and R_3 is the reading for door 3. Dilute the acetone extract with 90% acetone by a factor of 3 and by 10; measure the fluorescence for the more sensitive doors, 2 and 1, and determine $F_D{}^2$ and $F_D{}^1$ for each setting. Use the appropriate F_D value for each door setting in determining the amount of chlorophyll in an unknown sample.

Notes

(a) If a spectrofluorometer is available, the filters are obviously not necessary. Also, if a Turner Systems® fluorometer is used, the need to calibrate each door setting is not necessary (Section G).

(b) The chlorophyll concentration (in μg/ml) will be the chlorophyll contained in 10 ml of 90% extract. To express this as the chlorophyll in seawater, the value should be divided by the volume of seawater (in liters) and multiplied by the total volume of acetone extract (in ml). Note that the equations in Method 4.1 are for a 1-cm light path so that if 10-cm light path cuvettes are used to standardize this method, the μg/ml must be divided by 10; also, mg/m^3 $\equiv$ μg/l.

4.4. Fluorometric Determination of Phaeo-pigments

Introduction

The following procedure is taken from Holm-Hansen *et al.* (*J. Cons. perm. int. Explor. Mer*, **30**: 3, 1965). Using the equipment described for the fluorometric determination of chlorophyll *a*, the method has been standardized assuming a drop in fluorescent intensity of 2.2 before and after acid treatment of a chlorophyll extract, which did not previously contain phaeo-pigments (Note c).

Method

The procedure is similar to the previous method (4.3) except for the addition of acid to the acetone extract.

A. Experimental procedure

1. Extract chlorophyll from a water sample as described in the previous method (4.3) and obtain appropriate door factors for the instrument as described.
2. Measure the fluorescence of the sample before and after acidification. The acidification should be carried out with 2 drops of 5% v/v HCl and the second measurement made 30 sec after the first. The acid is mixed by inverting the fluorometer tube twice following addition (Note a).
3. Determine the amount of chlorophyll *a* and phaeo-pigments from the expressions,

$$\text{mg chlorophyll } a/\text{m}^3 = F_D \times 1.83(R_B - R_A)\frac{v}{V}$$

and

$$\text{mg phaeo-pigment/m}^3 = F_D \times 1.83(2.2R_A - R_B)\frac{v}{V}$$

where F_D, v and V are as determined in the previous method, and R_B and R_A are the fluorescence before and after the addition of acid, respectively (Note b).

Notes

(a) The tube must be thoroughly rinsed with 90% acetone to remove all trace of acid before the addition of the next sample.

(b) The method is not highly accurate because of the presence of other chlorophylls. Quantities of phaeo-pigments of less than 10% chlorophyll *a* should be interpreted with caution.

(c) If the ratio, R_B/R_A for pure chlorophyll, as measured using some other equipment is different from 2.2, then the factors in equations *A*3, above, become $(R_B/R_A)(R_B/R_A - 1)^{-1}$ instead of 1.83 in the first and second equations, and R_B/R_A instead of 2.2 in the second equation.

4.5. Automated Estimation of Chlorophyll Pigments

Introduction

This method is derived from Lorenzen (*Deep-Sea Res.*, **13**: 223, 1966). The method is neither highly accurate nor precise. However, it has the important advantage over all other methods in that it can be used to monitor chlorophyll concentrations while a vessel is under way, or in vertical profiles from pumped samples. Thus the ability to spot changes in concentration *in vivo*, as a survey is being conducted, is fundamental to the simultaneous collection of other kinds of data.

Method

A. Capabilities

The method is sensitive to < 0.1 mg chlorophyll a/m^3 depending to some extent on the instrument employed. However, the precision is generally $\pm 50\%$ over a wide range of samples taken both during the day and night. Precision can be improved over a small range of samples when there is less chance for change in species or physiological state of the algae.

B. Outline of method

Seawater is pumped directly through a Turner fluorometer, Model III (Note a), using a flow-through door, and the relative chlorophyll fluorescence is recorded on a strip chart recorder.

C. Special apparatus and equipment

A Turner Model III (Note a) fluorometer with a flow-through cell and cuvette (3/8 inch pipe); high intensity F4T5 blue lamp with a blue Corning filter CS.5–60 for excitation light and a red Corning filter CS.2–64 for the emitted light; photomultiplier R136, red sensitive.

A seawater system which allows rapid flow through the cuvette without building up high pressure such as might break the glass. A heat exchanger which warms the seawater to prevent condensation on the cuvette may be necessary in some cases.

A 10mV recorder to measure the output from the fluorometer.

D. Experimental procedure

1. Draw distilled water through the cuvette; turn off the flow-through system to conserve water and adjust the instrument to zero with each door setting.

2. Draw seawater through the system using a suitable door for the range of chlorophyll experienced.

3. Record changes in chlorophyll on the strip chart recorder as the vessel is underway, or over a vertical profile from a pumped sample.

4. An approximate measure of the chlorophyll content of the water can be obtained as

$$\text{mg chlorophyll } a/\text{m}^3 = F_D \times R$$

where F_D is determined for each door by the spectrophotometric method and R is the fluorometer reading.

E. Calibration

Filter a standard volume of seawater which has passed through the fluorometer and for which the fluorescence (R_S) is known (relative value in the range 10 to 90 units). Determine the chlorophyll content of this sample using Method 4.1. Then the factor F_D for any one door ($F_D{}^1$, $F_D{}^2$ and $F_D{}^3$) will be (Note b):

$$F_D = \frac{\text{mg chlorophyll } a/\text{m}^3}{R_S}$$

Notes

(a) The method is written up for a Model III Turner fluorometer but other types may be equally adaptable.

(b) Since the relative fluorescence of chlorophyll *in vivo* may change with species and physiological state of the algae, a number of estimates of F_D should be made at different fluorescent intensities and an average value obtained for any particular profile.

SECTION 5

Photosynthesis

5.1. Photosynthesis as Measured by the Uptake of Radioactive Carbon

Introduction

The following method is based on the radioactive carbon technique as originally described by Steemann Nielsen (*J. Cons. perm. int. Explor. Mer*, **18**: 117, 1952) and as modified for scintillation counting by Wolfe and Schelske (*J. Cons. perm. int. Explor. Mer*, **31**: 31, 1967). An alternative to the use of this technique is to arrange for the purchase of services through the International Agency for ^{14}C Determination. This agency will supply procedures, apparatus, ampules, and counting of samples; details may be obtained through Carbon 14 Centralen, Agern Allé 11, Dk-2970 Horsholm, Denmark.

The method described here does not take into account the interpretation of results. Many papers have been written both on the methodology of ^{14}C-uptake and on the interpretation of results. A review of these subjects is given by Carpenter and Lively in *Primary Productivity of the Sea*, Ed. P.G. Falkowski, *Publ.* Plenum Press, pp. 161–178, 1980.

Method

A. Capabilities

Range: 0.05–100 mg C/m³/hr

The range of this method depends very largely on the amount of radioactive carbon added and the precision of the radiochemical part of the procedure. There is virtually no upper limit and the lower limit could probably be reduced by using great care, particularly in respect to the dark blank.

Precision at the 30 mg C/m³/hr level:

The correct value lies in the range, mean of n determinations $\pm 3/n^{\frac{1}{2}}$ mg $C/M^3/hr$ (for a 3 hr incubation, 5 μCi added)

B. Outline of method

A known amount of radioactive carbonate, $^{14}CO_3^{2-}$, is added to a sample of seawater of known total carbonate content. After photosynthesis by the

CBMSA–E

endemic phytoplankton population has continued for a specific time period, the phytoplankton cells are filtered onto a membrane filter, washed, and the radioactivity from the carbon in the plants is measured with a suitable scintillation counter. The uptake of radioactive carbonate, as a fraction of the whole, is assumed to measure the uptake of total carbonate, as a fraction of the whole, and hence the rate of photosynthesis may be evaluated.

C. Special apparatus

BOD bottles, or similar bottles having a smaller volume and non-toxic screw-cap tops, cleaned with chromic-sulfuric acid and rinsed thoroughly with distilled water. Both clear and opaque bottles are required; the latter may be made from the former by masking with black tape.

Small sacks of black cloth and light-tight boxes to shield the sample bottles from light during the innoculation.

25-mm diameter Millipore (or equivalent) filtration apparatus, fitted with a funnel to hold the entire contents of the seawater bottle (i.e., 300 ml if BOD bottles are used). A manostat device should be used in the vacuum line to prevent the vacuum becoming greater than *ca.* 1/3 of an atmosphere.

Scintillation counter, preferably one with automatic sample selection and efficiency controls.

2-ml capacity syringe with extra long (10-cm) needles

Scintillation vials; 25-mm Millipore AA filters or equivalent; scintillation cocktail suitable for dissolving Millipore filters and with low quenching in the presence of small amounts of water.

D. Sampling procedure and storage

Sampling bottles should not contain any metal and are preferably teflon-lined. Incubations should be carried out immediately following the collection of samples. Subsurface samples should not be exposed to light during the filling operation or else "light shock" may alter the photosynthetic response of phytoplankton growing at depth in dim light.

E. Special reagents

1. Filtered seawater

As a wash liquid, use seawater taken at the same time as the samples and filter through an AA Millipore filter. Synthetic seawater or sodium chloride solution should not be used for washing filters.

2. Radioactive carbonate ampules

These can generally be purchased from scientific supply houses and should be either 1, 5, or 20 μCi Na$_2$ ^{14}CO$_3$ per ampule made up in 2 ml of 3.5% NaCl. If the ampules are purchased from the Carbon 14 Centralen, their stated activity can be assumed correct. Other ampules should be checked by diluting one ampule with 0.05N NaOH to give an approximate activity of 10,000 cpm/ml. Exactly 1 ml of this solution is then diluted with 9 ml of a scintillation fluid suitable for containing 1 ml of alkaline water, and the dpm measured using a scintillation counter, corrected for quenching.

3. Quaternary base and/or hydrogen peroxide

Some phytoplankton samples may be difficult to dissolve in the scintillation fluid. A few hours pre-treatment of samples with 1 ml of a quaternary base (available from supply houses) may be necessary in such cases. In addition, if the sample contains a large amount of chlorophyll a, this may interfere with counting and can be destroyed with 1 to 2 drops of 30% hydrogen peroxide.

4. Scintillation fluid

Many suitable scintillation fluids are now available for dissolving the filters. Among those mentioned in the literature are "PCS" from Nuclear Chicago, "Econofluor" or "Aquasol" from New England Nuclear, and "Insta-gel" from Packard Instrument Co. The quenching characteristics of the scintillation fluids in the presence of small amounts of water should be studied for the scintillation fluid chosen.

F. Experimental procedure

1. Seawater samples are poured into clear BOD (or equivalent) bottles, leaving an air space of about 3–5 ml at the top of the bottle (Note a).

2. A 2-ml ampule of radioactive carbonate is broken along its fracture zone (usually marked as a blue line), and the entire contents are transferred to the bottom of one sample bottle using a 2-ml syringe fitted with an extra long needle. A small amount of seawater is then withdrawn from the top of the sample bottle and this is used to rinse out the ampule, returning the rinsed seawater to the bottom of the sample bottles. Ampules should be rinsed at least twice (Note b).

3. Mix the contents of the sample bottles by shaking gently, cover with a black sack and place in a light-tight box. When all the samples have been innoculated they may be exposed to light, either in an incubator or by placing the bottles on a buoyed line for incubation *in situ*. Remove black bags from the bottles just prior to the start of incubation (Note c).

4. At the end of the incubation, replace sample bottles in black bags or in a light-tight box. Filter all samples as soon as possible, preferably using a multiple outlet filtering apparatus capable of simultaneously filtering up to 10 samples at once. Suction should not exceed 1/3 of an atmosphere during filtration. 25-mm AA Millipore filters may be used to collect the plankton and filters should be sucked dry and washed with 3 small aliquots (*ca.* 5 ml) of filtered seawater (Notes d and e).

5. Remove the filter from the holder while maintaining a vacuum and place in a scintillation vial containing 10 ml of scintillation fluid (Note f). Allow to stand overnight after shaking the sealed vial.

6. Count the vials in a scintillation counter and use the following equation to calculate photosynthesis (Note g):

$$\text{photosynthesis in mg } C/m^3/hr = \frac{(R_S - R_B) \times W}{R \times N}$$

where R is the total activity (dpm) of bicarbonate added (for exactly 5 μCi, this equals 1.11×10^7 dpm); N is the number of hours of incubation; R_S is the sample count (dpm) corrected for quenching (Note h); R_B is the dark bottle (or blank) count (dpm) corrected for quenching (Note h); and W is the weight of total carbon dioxide present in mg C/m^3 as determined from the expression, $W = 12,000 \times TC$, where TC is the total carbon dioxide as determined in Method 7.2. An approximation of the total carbon dioxide can be made for oceanic waters from the salinity as follows:

If the salinity, S‰, is known, then:
   Total alkalinity = S‰ × 0.067 meq/l
   Carbonate alkalinity = total alkalinity − 0.05
   Total carbon dioxide = 0.96 × carbonate alkalinity
Example: Salinity = 30.00‰
          Total alkalinity = 2.01 meq/l
          Carbonate alkalinity = 1.96 meq/l
          Total carbon dioxide = 1.88 meq/l
          W = 22,600 mg C/m^3

G. Determination of blank

Dark fixation of carbon dioxide may occur as a result of heterotrophic activity. However, the reason for the retention of radioactive material on a filter from a sample bottle which has been made completely opaque may be associated with a number of causes, some of which are not biological. These include retention of inorganic carbonate by the filters, radioactive particulate material as contaminants in the ampules and passive adsorption of carbonate by the filtered seston. In cases of relatively high photosynthesis, the dark blank is generally small (less than 5% of light fixation). However, in cases of low photosynthesis and particularly if there is extensive heterotrophic activity, dark fixation can be greater than 50% of the light bottle fixation. In such cases, a reason for high dark uptake should be investigated. As a general procedure, a dark blank should be run for each light bottle as follows:

Fill a blackened BOD bottle (or equivalent) with the sample and treat it exactly as described in Section F, 1–5, placing the dark bottle near to the light bottle (hence at the same temperature) during the illumination period. The final activity R_B (cpm) should be corrected for quenching to give disintegrations per minute (dpm).

H. Calibration

The standardization of radioactive material in commercially prepared ampules should be checked as described in Section E, 2. In addition, however, it has been found that some commercial ampules may contain small amounts of radioactive soluble organic material, particulate material and radioactivity other than ^{14}C. These contaminants can be removed by redistilling carbon dioxide from a large batch of radioactive carbonate and making up new ampules with the distillate. However, in general this will not be necessary, except where studies may be required on the filtrate containing organic material released during photosynthesis (Note i).

In some cases it has been reported that toxic materials, such as copper, may be present in the radioactive ampules. For this reason, it may be advisable to check the activity of photosynthetic experiments carried out with the researchers' ampules against those prepared by the Carbon 14 Centralen. Further, it is suggested that intercalibration of techniques with other scientists in the field should be more commonly practised in the particular case of ^{14}C-productivity measurements because of the lack of an external means of standardizing the entire procedure.

Notes

(a) An exact volume of seawater need not be measured into each bottle, providing the entire bottle contents are filtered at the end of the experiment. This is because any differences in volume between two samples is cancelled out in the uptake calculation by the same difference in concentration of the added radioactivity.

The 3–5 ml air space is to allow for the addition of the radioactive material without overflow.

(b) Radioactive carbonate is sometimes forced into the neck of the ampule above the fracture zone. Ampules should be inspected before opening and any excess fluid above the fracture zone should be tapped down into the ampule before opening.

The amount of radioactivity (1, 5 or 20 μCi) depends upon the amount of photosynthetic organisms present in the sample. Generally, 1 μCi will be sufficient for eutrophic waters while 20 μCi may be needed in oligotrophic waters.

(c) The time period of incubation is a matter of individual choice, but generally periods of between 2 and 24 hr are employed for specific reasons depending upon what individual scientists want to measure (e.g., see discussion by Hobson *et al.*, *J. Fish. Res. Bd. Canada*, **33**: 1715, 1976).

(d) When filtration is complete following 3 washings, it is advisable to hold each filter over fuming HCl for 2 min in order to remove traces of inorganic carbonate.

(e) Killing of samples at the end of an incubation using formaldehyde is not generally recommended since rupture of some cells occurs on death. Preferably, samples should be retained in the dark until filtered, and filter holders should be opaque to avoid additional photosynthesis during filtration.

(f) Prior digestion with a quaternary ammonium salt is optional before the addition of the scintillation fluid. Addition of hydrogen peroxide for bleaching can be made after the addition of scintillation fluid.

(g) There are two second order corrections which have not been applied to the equation. The first is that the isotope ^{14}C may be taken up more slowly than ^{12}C. The second is that W should represent the total carbon dioxide content of the water, which is the carbonate alkalinity times a factor. In previous descriptions of this method the first correction has been applied as an increase of 1.05 and the second, as a decrease of 0.95. Considering the overall accuracy of the method, these two corrections have been considered to cancel one another.

(h) Quenching may be corrected to give disintegrations per minute (dpm) from counts per minute (cpm) using any one of several standard procedures recommended for a particular scintillation counter. These include the channels ratio method and either internal or external standards.

(i) The filtrate from the radioactive carbonate incubation is sometimes used to study the release of organic substances by algae during photosynthesis. For this purpose, the inorganic carbonate has to be removed by purging acidified filtrates and measuring the activity remaining as follows:

Add 1 ml of 37% H_3PO_4 to 125 ml of filtrate and purge by vigorously bubbling carbon dioxide through the solution (air or nitrogen gas will be less effective as purgatives). Purging should continue for at least 20 min; a 10 ml sample of the filtrate can then be mixed with 10 ml of Insta-gel (Packard Inst. Co.) for counting.

SECTION 6

Bacteria

6.1. Direct Counting of Bacteria by Fluorescence Microscopy

Introduction

Bacterial numbers in seawater may be estimated directly from counts when a fluorescent dye is used to clearly identify the bacteria. This method is greatly superior to early methods involving phase contrast counts. The method was originally described by Francisco *et al.* (*Trans. Am. Microsc. Soc.*, **92**: 416, 1973) and has been modified by Daley and Hobbie (*Limnol. Oceanogr.*, **20**: 875, 1975) and Hobbie *et al* (*Appl. Environ. Microbiol.*, **33**: 1225, 1977).

Method

A. Capabilities

The range of bacterial numbers that can be counted by this method depends on the quantity of seawater filtered. For bacterial numbers of 10^3 to 10^8/ml, the quantity of water can be readily adjusted to allow for suitable counts (Note a). The precision of counts depends to some extent on individual observers, but the following precision is given as a general guide:

Precision at the 10×10^5 cells/ml level:

The correct value lies in the range, mean of n determinations $\pm 2 \times 10^5/n^{\frac{1}{2}}$ cells/ml

B. Outline of method

A seawater sample is filtered onto a Nuclepore filter which has been previously stained black. The bacteria are stained with acridine orange, the filter is oiled, and bacteria, which fluoresce green against a black background, are counted in random fields (Note b).

C. Special apparatus and equipment

Millipore filters of 0.45 μ and 0.2 μ and pore size, 2.5 cm diameter
Nuclepore filters of 0.2 μ and 2.5 cm diameter
Acridine orange
Irgalan black

1 and 5-ml automatic pipettes

Millipore filtering apparatus with 2.5 cm fritted glass disc and *ca.* 15 ml tower reservoir

Zeiss Standard 18 microscope fitted with an IV FL epifluorescence condensor, a 100 W halogen lamp, a 455–500 band-pass filter, a 510 beam splitter and an LP528 barrier filter. (*Note:* a similar system can be purchased from Leitz)

Immersion oil, coverslips and microscope slides

D. Sampling procedure and storage

Samples should be collected in sterile Niskin bottles or Van Dorn bottles that have been well cleaned and rinsed with 70% ethyl alcohol. About 10 ml of seawater should be transferred to a screw-cap vial (Note c) and 0.5 ml of 40% formaldehyde added with a syringe. Mix and store in a cool, dark place (e.g., a refrigerator at 10°C). Samples should be analyzed within 2 weeks. The addition of formaldehyde is not necessary if the samples are going to be analyzed immediately (Note h).

E. Special reagents

1. Irgalan black (Note d)

Dissolve 2 g of irgalan black in 1 l of 2% acetic acid. The solution is stable indefinitely.

2. Acridine Orange

Dissolve 0.1 g of Acridine Orange (Chemical Index 46005) in 100 ml of distilled water. The dye does not dissolve readily and the reagent should be put on a flask shaker for *ca.* 1 hr. The reagent is filtered through a 0.45 μ Millipore filter and then through a 0.2 μ Millipore filter to remove particles of dye and any bacteria. The solution can be stored for several weeks in a refrigerator at 5°C. Refiltration of the solution may be necessary if high blank values are experienced.

3. Membrane filtered rinse water

Distilled water is filtered through a 0.2 μ Millipore filter to remove all bacteria. The rinse water should be filtered each time an analysis is performed.

4. Stained Nuclepore filters

Nuclepore filters (0.2 μ pore size, 2.5 cm diameter) are stained by soaking for 4 hr in a petri dish containing the irgalan black reagent. Filters are rinsed by dipping several times in the reagent rinse water and can be used immediately, or stored by drying on absorbant paper.

5. *Formaldehyde*

40% formaldehyde is filtered through a 0.2 μ Millipore filter.

F. *Experimental procedure*

1. Place a 0.45 μ Millipore backing filter onto the fritted glass of the Millipore filter assembly and place a stained Nuclepore filter, shiny side up, on top.

2. Assemble the Millipore filtering apparatus, securing the tower with a clamp.

3. Rinse a 5-ml automatic pipette with sample and dispense 2 ml of the sample into the filter tower. Add 0.2 ml of Acridine orange stain to the sample and mix by tilting the equipment gently. Incubate for exactly 2 min and then apply vacuum ($\frac{1}{2}$ atm) by starting the filtration pump. Draw the filter dry.

4. Remove the tower and place the damp filter on a glass slide (labelled). Add one drop of immersion oil (Note f) to the center of the filter without allowing the oil wand to touch the filter. Place a cover slip on top of the oil and filter, and press lightly to spread the oil and remove air bubbles.

5. The slide may be counted immediately or kept in the dark for no longer than 24 hr. Using the oil immersion objective, lower the objective lens until it just touches a drop of immersion oil placed above the coverslip. Turn on the microscope lamp and adjust voltage to the recommended optimum. Looking through the ocular, slowly focus the microscope onto the bacteria.

6. Count at least ten fields (Note e) with about 20 bacteria per field.

7. Calculate the number of bacteria in the sample as the mean ($\bar{x}$) of ten field counts (corrected for blank):

$$\text{cells/ml} = \bar{x} \times \frac{\text{filter area}}{\text{field area}} \times \frac{1}{\text{ml filtered}}$$

where the filter area is the area of the entire filter surface on which the bacteria were collected (about $2 \times 10^8 \ \mu^2$ for a 2.5 cm filter) and the field area at X1000 magnification is the area in which the bacteria were counted (for a whole graticule at X1000 this usually lies in the range 10^3 to $10^4 \ \mu^2$ and should be measured precisely). The ml of water filtered in Step 3 above was 2.

8. Rinse all filtering apparatus with the reagent rinse water before proceeding to the next sample in order to avoid contaminating subsequent counts.

G. Determination of blank

Carry out the procedure in Section F, 1–7, using a 2-ml sample of rinse water containing formaldehyde. Correct the mean ($\bar{x}$) of the sample counts by the mean of three blank determinations.

H. Calibration

There is no direct calibration of the method, but the statistics of counting should be checked by each individual using the technique (Note g).

Notes

(a) Suitable counts are made when the number of bacteria in the eyepiece graticule (ca. 100 μ diameter) of the microscope is between about 15 and 30. If dilution of the seawater sample is necessary, 0.2 μ-filtered seawater must be used to dilute the sample.

(b) The intensity and color of the fluorescence is dependent to some extent on the biochemical composition of different bacteria and the incubation time with the stain.

(c) Scintillation vials make suitable sample containers.

(d) Irgalan black (Chemical Index, acid black 107) is available from Ciba-Geigy Corp., Dyestuffs and Chemical Division, Greensboro, N.C. and from Union Color and Chemical, Boston, Mass.

(e) The graticule should have subdivisions so that if bacteria are very numerous, then a "field" can be 0.1 or 0.25 of the total graticule area.

(f) The immersion oil must have a low fluorescence.

(g) An alternate stain to Acridine Orange is reported to improve the accuracy of counting. The stain is 4'6-diamidino-2-phenylindole (DAPI) and the technique is reported by Porter and Feig (*Limnol. Oceanogr.*, **25**: 943, 1980).

(h) If 0.5 ml of formaldehyde is added, then the number of bacteria should be multiplied by 1.05 in Section F, 7.

6.2. Heterotrophic Activity (as Measured by Glucose Uptake)

Introduction

The following method is designed to measure heterotrophic activity through a determination of the amount of added organic substrate taken up in a unit of time. The original method was described by Parsons and Strickland (*Deep-Sea Res.*, **8**: 211, 1962), and it was further investigated by Wright and Hobbie (*Ecology*, **47**: 447, 1966). These two references show that by the use of a radioactive organic substrate, the maximum velocity of uptake (V_{max}) for a population of heterotrophs can be measured and interpreted in terms of both the size and metabolic activity of the population. Since practically all the heterotrophic activity can be attributed to bacteria, a separate measure of the size of the population (e.g., Method 6.1) will allow for the measurement of V_{max} to be interpreted as the specific activity (i.e., uptake activity, V_{max}, divided by the standing stock of bacteria).

The calculation of V_{max} requires serial dilutions of substrate and is therefore rather tedious to carry out as a routine method. In a more recent paper, Griffiths *et al.* (*App. Environ. Microbiol.*, **34**: 801, 1977) have shown that, providing sufficient substrate is added to the seawater, the measured uptake is closely correlated with V_{max} (r = 0.8 to 0.9). For glucose, this rate saturating concentration is in the range of 50 to 100 $\mu g/l$; for glutamic acid, it is considerably less, *ca.* 1 to 10 $\mu g/l$.

For further interpretation of results obtained with this method, the reader should refer to the references cited as well as to other more recent papers giving actual results.

Method

A. Capabilities

Range: 0.01 to 1.0 μg glucose/l/hr (Note a)
Precision at the 0.05 μg glucose/l/hr level:
The correct value lies in the range, mean of n determinations $\pm 0.006/n^{\frac{1}{2}}$ μg glucose/l/hr

B. Outline of method

D-[U-^{14}C] glucose is added to seawater in the presence of excess glucose substrate. The sample is incubated and filtered after a short time period. The amount of radioactive glucose retained on the filter is used as a measure of heterotrophic activity.

C. Special apparatus

125-ml screw-cap bottles, cleaned with acid-chromate and well rinsed in distilled water. (Bottles should either be painted black or incubations carried out in the dark.)
Millipore filtering apparatus for 25 mm, HA Millipore filters

Scintillation vials and scintillation counter
Automatic pipette for innoculating samples; wash bottle

D. Sampling procedure and storage

Heterotrophic activity measurement should be carried out only on freshly collected samples in order to avoid any change in bacterial populations in bottles. Samples should be innoculated with substrate and incubated immediately. Washed filters in scintillation vials, to which the scintillation cocktail has been added, can be stored indefinitely.

E. Special reagents

1. Radioactive glucose

D-[U-^{14}C] glucose having an activity of 5 μCi/ml (Note b) is made up with 10 μg of inactive glucose/ml of 3.5% saline. The solution is sealed in ampules (usually in 10-ml batches) and autoclaved at 15 psi for 15 min. Each ml of this solution, when pipetted into 125 ml of seawater, will give a total activity of 5 μCi and a glucose concentration of 80 μg/l (Note e).

2. Filtered seawater

Seawater required for washing the filtered samples should be prepared by filtering water from the same area as the experiments using a HA Millipore filter and retaining about 500 ml of filtrate in a wash bottle.

3. Scintillation fluid ("cocktail")

A number of "cocktails" are commercially available. These include "PCS" from Nuclear Chicago, "Econofluor" or "Aquasol" from New England Nuclear and "Insta-gel" from Packard Instrument Co.

F. Experimental procedure

1. Seawater samples are poured into 125-ml screw-cap bottles leaving a small air space for the innoculum (Note c).
2. Add 1 ml of radioactive glucose solution to each bottle, stopper and mix gently.
3. Incubate for 1 hr at *in situ* temperature (Note d).
4. Filter all samples simultaneously on a multiple filter outlet at 1/3 atm. Rinse the filters three times with filtered seawater (*ca.* 5 ml per rinse). Remove the filter while still maintaining the vacuum.

5. Place each filter in a separate scintillation vial, add 10 ml of scintillation cocktail, and count after 24 hr on a scintillation counter for ^{14}C.

6. Correct the count for quenching to convert counts per minute (cpm) into disintegrations per minute (dpm) and determine the glucose uptake as:

$$\mu g \text{ glucose/l/hr} = \frac{(R_S - R_B) \times A}{C \times t}$$

where A is the concentration of added substrate (80 μg/l in this example; Note e), R_S is the dpm of the sample, R_B is the dpm of the blank, C is the total radioactivity added (1.11×10^7 dpm in this example), and t is the time of the incubation (usually 1 hr).

G. Determination of blank

Add 1 ml of radioactive glucose to a 125 ml sample of seawater, mix and filter immediately. Carry out steps 4 to 6, Section F, and find R_B for each sample.

H. Calibration

The method depends on accurately knowing the activity of the added glucose. If this is not given at the time of purchase, a serial dilution of a stock solution must be made and the dpm of the glucose determined.

Notes

(a) The method is written up for determining the uptake activity of glucose. The uptake activity for amino acids, especially glutamic acid, is generally higher while for acetate it is generally lower, than for glucose. The range over which measurements can be made will also depend on the specific radioactivity of the added substrate.

(b) The actual amount of radioactive glucose added should be in the range 1 to 10 μCi for most seawater samples. Its absolute activity in dpm must be known for each ml added to the seawater samples.

(c) The 125 ml sample does not have to be highly precise (i.e., ± 3 ml) since small changes in concentration of substrate are cancelled out by small changes in total volume of water filtered.

(d) The incubation time should be as short as possible in order not to have an appreciable increase in bacterial numbers, while still allowing time for sufficient radioactivity to be taken up.

(e) It is assumed that the inactive glucose is a large excess over the radioactive glucose added. If this is not the case, then the concentration of radioactive glucose must be added to the value A (i.e. at specific activities of < 500 mCi/mM).

6.3. Heterotrophic Growth
(as Measured by Thymidine Uptake)

Introduction

In the previous method (6.2) on heterotrophic activity, the property measured was largely metabolic activity, either of growth and/or basal metabolism. The following method is believed to be specific for measuring heterotrophic growth on the premise that only growing cells incorporate radioactive thymidine into deoxyribonucleic acid (DNA). As in the previous method, this incorporation is believed to be largely bacterial so that the method essentially measures *bacterial growth.* The method is that given by Fuhrman and Azam (*App. Environ. Microbiol.*, **39**: 1085, 1980 and *Mar. Biol.*, **66**: 109, 1982). In the latter reference, the authors provide an approximate conversion factor from a regression of log-transformed changes in bacterial numbers versus log-thymidine incorporation ($r^2 = 0.69$; $n = 11$); the factor reported is 1.4×10^{18} cells produced per mole of thymidine incorporated.

Method

A. Capabilities

Range: 1×10^5 to 1×10^9 cells/l/day
Precision of this method should be determined experimentally on replicate samples.

B. Outline of method

The method assumes that all growing bacterial cells will incorporate radioactive thymidine added to seawater into deoxyribonucleic acid. The amount of thymidine incorporated is then taken as a measure of growth. The method also assumes that the amount of DNA required by a bacterial cell for growth is constant (i.e., *ca.* 2.6×10^{-15}g of DNA per bacterium); that the amount of natural thymidine does not affect the uptake of radioactive thymidine; and, that the uptake of thymidine by phytoplankton is insignificant. A discussion of these assumptions is given in the literature cited above.

C. Special apparatus

25-mm, 0.45 μ Millipore filters and filtering apparatus (Note a)
50-ml plastic test tubes stoppered with cotton wool
Scintillation vials; scintillation counter

D. Sampling procedure and storage

Samples should be analyzed immediately following collection. Samples can be collected in sterile Niskin bottles or well cleaned Van Dorn bottles. Samples may be stored only after they have been placed in the scintillation fluid.

E. Special reagents

1. Radioactive thymidine

Thymidine [methyl-^{3}H] of high specific activity (> 50 Ci/mmol) purchased from New England Nuclear (Boston) or ICN (Irvine, California) or equivalent supplier. Store refrigerated in 70% aqueous ethanol for maximum stability. For experimental use, the solution is evaporated to dryness with a stream of filtered air and is reconstituted in filtered, sterilized, distilled water to a final concentration of 1 mCi/ml.

2. Trichloroacetic acid (TCA)

(i) Make a 10% extraction solution of TCA with distilled water.
(ii) Make a 5% rinse solution of TCA with distilled water.

3. Ethyl acetate (reagent grade)

4. Scintillation fluid

(See Method 6.2 for a variety of cocktails)

F. Experimental procedure

1. Pour a 20 ml aliquot of seawater into a 50-ml plastic test tube. Add 0.25 ml of a solution containing 1 mCi/ml thymidine (i.e., 5 nM, or 250 μCi of a 50 Ci/mmol thymidine preparation) and mix by swirling.
2. Incubate for 30 min at *in situ* temperature (Note b).
3. After incubation, the samples are poured into plastic test tubes that have been chilled in an ice bath. After 1 min, 20 ml of ice cold, 10% TCA is added; mix and allow to extract for 5 min.
4. Collect the insoluble material by filtration onto a 0.45 μ pore-sized, 25-mm diameter Millipore filter, rinse 5 times with 1 ml portions of 5% TCA, and place the filter in a scintillation vial.
5. Add 1 ml of ethyl acetate to dissolve the filter (*ca.* 10 min required) and add 10 ml of scintillation fluid (e.g., Aquasol). The vial is placed in a

scintillation counter and the dpm determined from cpm, corrected for quenching (Note c) and the blank.

6. The number of moles of thymidine incorporated is given by:

$$\text{mmoles/l/hr} = \frac{U}{S} \times \frac{4.5 \times 10^{-13}}{t} \times \frac{1}{v}$$

where U is the dpm of the filter times 4.5×10^{-13} which is the number of curies per dpm, S is the specific activity in Ci/mmole, t is the time of the incubation in hours, and v is the volume of sample incubated in liters. The conversion of thymidine incorporated to bacterial number can be carried out using a conversion factor as suggested in the Introduction (Note d).

G. Determination of blank

To a 20 ml aliquot of seawater, add 0.1 ml of 40% formalin from an automatic pipette. Carry out the procedure 1 to 5 in Section F and subtract the blank scintillation count (in dpm) from the unknown sample.

Notes

(a) $0.22\ \mu$ pore size filters may be needed for deep ocean samples. The difference in pore size effectiveness versus ease of filtration may have to be determined experimentally.

(b) The length of the incubation will depend on the seawater sample. For coastal areas, it may be necessary to filter after < 30 min, while for some open ocean and deep water samples, incubations may have to continue for *ca.* 3 hr to obtain a sufficient count.

(c) Any quenching correction is probably best determined by having an internal ^{3}H-toluene standard on a duplicate sample.

(d) Thymidine synthesis by bacteria during the course of the incubation will yield low growth estimates. The problem of calibrating the method to actual growth conditions is discussed by Kirchman *et al.* (*Appl. Environ. Microbiol.*, **44**: 1296, 1982).

SECTION 7

Gases in Seawater

7.1. Determination of Dissolved Oxygen

Introduction

The method described here follows a modification of the classical Winkler procedure. A discussion of the accuracy of the technique and various modifications has been given by Carpenter (*Limnol. Oceanogr.*, **10**: 141, 1965) and Carritt and Carpenter (*J. Mar. Res.*, **24**: 286, 1966). The latter reference gives a procedure for the absolute calibration of the Winkler method by titration. In the following procedure, a spectrophotometric method for oxygen analysis is employed as being more rapid and requiring less procedural standardization without a loss of precision. It is also a particularly sensitive method at oxygen concentrations of < 1 ppm. However, the spectrophotometric method lacks an absolute standard other than an assumed air-saturated standard. Since oxygen supersaturation can occur, including during some *in vivo* phytoplankton blooms, the experimenter must assure that the calibration factor is correct and, if necessary, check the oxygen content of several standards by using the titrametric method (Note c). The spectrophotometric method used here follows that given by Duval *et al.* (*J. Fish. Res. Bd. Canada*, **31**: 1529, 1974), and the oxygen saturation values (Table 2) are from Weiss (*Deep-Sea Res.*, **17**: 721, 1970).

Method

A. Capabilities

Range: 0.1 to 10 mg O_2/l
Precision at the 9.5 mg/l level:
  The correct value lies in the range, mean of n determinations $\pm\, 0.064/n^{\frac{1}{2}}$ mg/l

B. Outline of method

A divalent manganese solution, followed by strong alkali, is added to the sample. The precipitated manganous hydroxide is dispersed evenly throughout the seawater sample which completely fills a stoppered glass bottle. Any dissolved oxygen rapidly oxidizes an equivalent amount of divalent manga-

135

nese to basic hydroxides of higher valency states. When the solution is acidified in the presence of iodide, the oxidized manganese again reverts to the divalent state and iodine, equivalent to the original dissolved oxygen content of the water, is liberated. The amount of iodine is measured spectrophotometrically.

C. Special apparatus and equipment

125-ml BOD or glass-stoppered Erlenmeyer flasks
Automatic syringe pipettes
Spectrophotometer capable of reading in the UV range (*ca.* 280 nm)

D. Sampling procedure and storage

Clean BOD bottles should be rinsed twice with seawater before sampling. Samples should be introduced through a long tube into the bottom of the bottle so as to avoid bubbles while filling; the seawater should be allowed to overflow the bottle before stoppering. Two reagents, the manganous chloride and the alkaline iodide, should be added immediately and the contents of each bottle mixed by inverting. The bottles should be kept in a dark box at the same temperature and preferably analyzed within the next 6 hr.
Caution: This procedure employs a strong acid and alkali.

E. Special reagents

1. Manganous chloride (3M)

Dissolve 600 g $MnCl_2 \cdot 4H_2O$ in distilled water and make to a volume of 1 l.

2. Sodium hydroxide (8N)/Iodide (3M)

Dissolve 320 g NaOH and 600 g NaI in distilled water and make to a volume of 1 l after allowing to cool to room temperature.

3. Sulfuric acid (10N)

Add 280 ml of concentrated H_2SO_4 to about 500 ml of distilled water; add slowly with stirring. Allow to cool and make up to a volume of 1 l with distilled water.

F. Experimental procedure

1. Remove the stopper of the BOD bottle and add 1 ml of the manganous

reagent and 1 ml of sodium iodide/hydroxide reagent using a syringe pipette (Note a). Stopper firmly and mix by inversion. Allow the precipitate to settle half way down the bottle.

2. Add 1 ml of the sulfuric acid reagent, stopper and mix thoroughly (Note b). All the precipitate should redissolve after the addition of acid.

3. 3 to 5 min after acidification, a 5 ml aliquot of the iodine-containing solution is diluted to 100 ml with distilled water; care is taken in mixing (to minimize oxidation of iodine through contact with air) by inverting the flask gently.

4. Measure the extinction of the iodine (E_R) at 287.5 nm using a 1-cm quartz cell in a UV spectrophotometer. Then,

$$mg\ O_2/l\ (ppm) = F_S\ (E_R - E_B)$$

where F_S and E_B are determined as described below.

G. Determination of blank

Carry through the procedure 1 to 4, Section F, using water through which nitrogen has been bubbled to remove all the oxygen. Measure the extinction (E_B) at 287.5 nm.

H. Calibration

The calibration factor in this procedure may have to be obtained against the standard titration technique (see Introduction). Alternatively, if air is bubbled into seawater of known salinity and temperature at sea level, Table 2 can be interpolated to determine the oxygen content of the water. This sample is then taken through the procedure F, steps 1 to 4, to give an extinction, E_S. Then the factor F_S, is given as:

$$F_S = \frac{\text{concentration of oxygen in mg/l}}{E_S - E_B}$$

where F_S should be close to 16 for the conditions of the reaction (including the 1:20 dilution of the sample before measuring the extintion; Note c).

Notes

(a) Place the tip of the pipette just below the surface of the seawater sample. The heavy reagents sink. The two reagents should be added immediately and mixed after both have been added.

(b) Small volume changes due to the addition of reagents are the same for standard and sample and are not therefore corrected independently.

Gases in Seawater

TABLE 2. Oxygen solubility in cm³/dm³ (= ml/l) of air in seawater at an atmospheric pressure of 760 mm Hg calculated according to Weiss
(*Deep-Sea Res.*, **17**: 721, 1970)

T (°C)	0	2	4	6	8	10	12	14	16	18	20	22	24	26	28	30	32	34
									S (‰)									
0	10.22	10.08	9.94	9.81	9.67	9.54	9.41	9.29	9.16	9.04	8.91	8.79	8.67	8.56	8.44	8.32	8.21	8.10
1	9.94	9.80	9.67	9.54	9.41	9.28	9.16	9.04	8.91	8.79	8.68	8.56	8.44	8.33	8.22	8.11	8.00	7.89
2	9.67	9.54	9.41	9.28	9.16	9.04	8.92	8.80	8.68	8.56	8.45	8.34	8.22	8.11	8.01	7.90	7.79	7.69
3	9.41	9.28	9.16	9.04	8.92	8.80	8.68	8.57	8.45	8.34	8.23	8.12	8.01	7.91	7.80	7.70	7.60	7.50
4	9.16	9.04	8.92	8.81	8.69	8.57	8.46	8.35	8.24	8.13	8.02	7.92	7.81	7.71	7.61	7.51	7.41	7.31
5	8.93	8.81	8.70	8.58	8.47	8.36	8.25	8.14	8.03	7.93	7.83	7.72	7.62	7.52	7.42	7.33	7.23	7.14
6	8.70	8.59	8.48	8.37	8.26	8.15	8.05	7.94	7.84	7.74	7.64	7.54	7.44	7.34	7.25	7.15	7.06	6.97
7	8.49	8.38	8.27	8.16	8.06	7.95	7.85	7.75	7.65	7.55	7.45	7.36	7.26	7.17	7.08	6.98	6.89	6.81
8	8.28	8.17	8.07	7.97	7.86	7.76	7.66	7.57	7.47	7.37	7.28	7.19	7.09	7.00	6.91	6.82	6.74	6.65
9	8.08	7.98	7.88	7.78	7.68	7.58	7.48	7.39	7.30	7.20	7.11	7.02	6.93	6.84	6.76	6.67	6.59	6.50
10	7.89	7.79	7.69	7.60	7.50	7.41	7.31	7.22	7.13	7.04	6.95	6.86	6.78	6.69	6.61	6.52	6.44	6.36
11	7.71	7.61	7.52	7.42	7.33	7.24	7.15	7.06	6.97	6.88	6.80	6.71	6.63	6.54	6.46	6.38	6.30	6.22
12	7.53	7.44	7.35	7.26	7.17	7.08	6.99	6.90	6.82	6.73	6.65	6.56	6.48	6.40	6.32	6.24	6.17	6.09
13	7.37	7.27	7.18	7.10	7.01	6.92	6.84	6.75	6.67	6.59	6.50	6.42	6.34	6.27	6.19	6.11	6.04	5.96
14	7.20	7.12	7.03	6.94	6.86	6.77	6.69	6.61	6.53	6.45	6.37	6.29	6.21	6.14	6.06	5.99	5.91	5.84
15	7.05	6.96	6.88	6.79	6.71	6.63	6.55	6.47	6.39	6.31	6.24	6.16	6.08	6.01	5.94	5.87	5.79	5.72
16	6.90	6.81	6.73	6.65	6.57	6.49	6.41	6.34	6.26	6.18	6.11	6.03	5.96	5.89	5.82	5.75	5.68	5.61
17	6.75	6.67	6.59	6.51	6.44	6.36	6.28	6.21	6.13	6.06	5.99	5.91	5.84	5.77	5.70	5.64	5.57	5.50
18	6.61	6.54	6.46	6.38	6.31	6.23	6.16	6.08	6.01	5.94	5.87	5.80	5.73	5.66	5.59	5.53	5.46	5.40
19	6.48	6.40	6.33	6.25	6.18	6.11	6.03	5.96	5.89	5.82	5.75	5.69	5.62	5.55	5.49	5.42	5.36	5.29
20	6.35	6.28	6.20	6.13	6.06	5.99	5.92	5.85	5.78	5.71	5.64	5.58	5.51	5.45	5.38	5.32	5.26	5.20
21	6.23	6.15	6.08	6.01	5.94	5.87	5.80	5.74	5.67	5.60	5.54	5.47	5.41	5.35	5.28	5.22	5.16	5.10
22	6.11	6.04	5.97	5.90	5.83	5.76	5.69	5.63	5.56	5.50	5.44	5.37	5.31	5.25	5.19	5.13	5.07	5.01
23	5.99	5.92	5.85	5.79	5.72	5.65	5.59	5.52	5.46	5.40	5.34	5.28	5.21	5.15	5.10	5.04	4.98	4.92
24	5.88	5.81	5.74	5.68	5.61	5.55	5.49	5.42	5.36	5.30	5.24	5.18	5.12	5.06	5.01	4.95	4.89	4.84
25	5.77	5.70	5.64	5.58	5.51	5.45	5.39	5.33	5.27	5.21	5.15	5.09	5.03	4.98	4.92	4.86	4.81	4.75
26	5.66	5.60	5.54	5.48	5.41	5.35	5.29	5.23	5.17	5.12	5.06	5.00	4.95	4.89	4.83	4.78	4.73	4.67
27	5.56	5.50	5.44	5.38	5.32	5.26	5.20	5.14	5.08	5.03	4.97	4.92	4.86	4.81	4.75	4.70	4.65	4.60
28	5.46	5.40	5.34	5.28	5.23	5.17	5.11	5.05	5.00	4.94	4.89	4.83	4.78	4.73	4.67	4.62	4.57	4.52
29	5.37	5.31	5.25	5.19	5.14	5.08	5.02	4.97	4.91	4.86	4.81	4.75	4.70	4.65	4.60	4.55	4.50	4.45
30	5.28	5.22	5.16	5.10	5.05	4.99	4.94	4.89	4.83	4.78	4.73	4.68	4.62	4.57	4.52	4.47	4.43	4.38

(c) The calibration factor depends to some extent on the quality of reagents used, the dilution procedure, and an individual's use of reagents. Users should develop their own F_S value which should be reproducible.

Addendum

1. Increase in precision

The precision of the Winkler oxygen technique, as determined by titration of the iodine, can be improved by a factor of *ca.* 10 if a photometric detector is used for the end point (Bryan *et al., J. exp. mar. Biol. Ecol.*, **21**: 191, 1976). This allows for the technique to be used in photosynthetic production studies as an alternative to the highly sensitive ^{14}C-technique (Method 5). However, for routine determination of the oxygen content of seawater, the spectrophotometric method described here is sufficient.

2. Interfering substances

Any substances with oxidizing or reducing potential present in seawater, other than oxygen, are possible sources of interference in the Winkler oxygen technique. These include nitrite, ferrous iron and organic matter (such as may come from very high densities of phytoplankton or, in coastal regions, from sewage, land drainage or industrial outfalls). These problems are not general, but special precautions may be necessary if the experimenter has any reason to suspect the results obtained by the method given above. Some reference material that may be useful is as follows:

Interfering substance	*Reference*
Nitrite	Alsterberg (*Biochem. Z.*, **159**: 36, 1925)
Iron	Alsterberg (*Biochem. Z.*, *170*: 30, 1926)
Organic matter	Pomeroy and Kirschman (*Industr. Eng. Chem.* (Anal.), **17**: 715, 1945)
Modifications in procedure	American Publ. Health Assoc. Standard Methods. N.Y. 13th Ed. Pp. 477–484.

7.1.1. Oxygen Standardization
by Titration

Introduction

The following technique should be used to standardize the spectrophotometric method for oxygen determination. The method also can be used for the
routine determination of oxygen in seawater, but as such it is more time
consuming than the spectrophotometric method.

Method

A. Outline of method

The iodine liberated using Method 7.1 is titrated with standardized
thiosulfate solution to give the absolute amount of oxygen in the original
sample. The method is taken from Carpenter (*Limnol. Oceanogr.*, **10**: 141,
1965) and Carritt and Carpenter (*J. Mar. Res.*, **24**: 286, 1966).

B. Special apparatus and equipment

In addition to apparatus described in Method 7.1:
50-ml pipette and 10-ml burette, both automatic zero adjusted
125-ml Erlenmeyer flask painted white on one side
Magnetic stirring bar and stirrer
Good illumination from a fluorescent lamp for end point detection

C. Special reagents

1. *Reagents as in Method 7.1.*
2. *Standard thiosulfate solution (ca. 0.01N)*
   Dissolve 2.9 g of analytical grade sodium thiosulfate, $Na_2S_2O_3 \cdot 5H_2O$,
and 0.1 g of sodium carbonate, Na_2CO_3, in 1 l of distilled water. Add 1 drop of
carbon bisulfide as a preservative.
3. *Standard potassium iodate (0.010N)*
   Dry analytical quality potassium iodate, KIO_3, at 105°C for 1 hr; cool
and weigh out exactly 0.3467 g. Dissolve in 200–300 ml of distilled water,
warming if necessary; cool and make up to 1 l with distilled water.
4. *Starch indicator*
   A 1% solution of starch: dissolve in dilute NaOH and neutralize with
dilute HCl to aid solution if necessary.

D. Experimental procedure

1. Within 1 hr of the acidification of the sample(s) in Method 7.1, transfer 50.0 ml into the specially painted Erlenmeyer flask using the self-adjusting pipette. Titrate immediately with 0.01N thiosulfate contained in the burette; stir with the magnetic stirrer until the solution is faintly yellow.

2. Add 0.5 ml of starch solution and continue the titration cautiously until the blue color disappears. Subtract the blank (Section G, Method 7.1) to obtain the corrected titration volume, V in ml, and calculate the oxygen content in mg O_2/l as:

$$\text{mg } O_2/l = 0.1016 \times f \times V \times 16$$

when a 50 ml aliquot is taken from a 125-ml BOD bottle and where f is determined as described below.

E. Calibration of thiosulfate

1. To a 125-ml BOD bottle filled with distilled water add 1 ml of concentrated sulfuric acid and 1 ml of alkaline iodide solution; mix thoroughly. Add 1 ml of the manganous chloride solution and mix again (i.e., reverse order of reagent addition in Method 7.1, Section F).

2. Withdraw 50 ml into the painted titration flask; add 5.0 ml of 0.01N standard iodate. Mix gently for 2 min and titrate with thiosulfate as above (Section D, 1 and 2). If V is the titration volume, then

$$f = \frac{5.00}{V}$$

where f should be the mean for at least 3 replicates.

C. Special apparatus and equipment

A highly sensitive pH meter with an expanded scale for accurate readings 200-ml, wide mouth, screw-capped polyethylene bottles to collect samples

D. Sampling procedure and storage

Samples should be collected in clean 200-ml, wide mouth, screw-capped polyethylene bottles and analyzed within a few hours of collection.

E. Special reagents

1. Standard 0.01N hydrochloric acid

1.0N acid can be purchased from chemical supply houses and should be diluted one in a hundred.

2. Standard buffer (pH 4.00 at 20–25°C)

Dissolve 10.21 g of analytical grade potassium hydrogen phthalate, $KHC_8H_4O_4$, in distilled water and make to a volume of 1000 ml. Store in a glass bottle.

3. Standard buffer (pH 6.87 at 20–25°C)

Dissolve 34.0 g of potassium dihydrogen phosphate, KH_2PO_4, and 35.5 g of anhydrous disodium hydrogen phosphate, Na_2HPO_4, in distilled water and make to a volume of 1000 ml. Store in a tightly stoppered glass bottle.

Dilute 100 ml of this buffer to 1000 ml with distilled water for use. The dilute solution can be preserved for several weeks with a few drops of chloroform.

7.2. Determination of Carbonate Alkalinity and Total Carbon Dioxide (all Forms)

Introduction

The method suggested here is one of several available involving the titration of carbonate in seawater. It is probably the simplest of these and is accurate for most seawater samples. The method is reproduced here essentially as described by Anderson and Robinson (*Ind. Eng. Chem. Anal. Ed.*, **18**: 767, 1946) for total alkalinity with a borate correction table determined by Dr. E.V. Grill (Dept. of Oceanography, University of British Columbia, Vancouver, Canada) using K'_B derived from Edmond and Gieskes (*Geochim. Cosmochim. Acta*, **34**: 1261, 1970).

Method

A. Capabilities

> *Range: 0.5 to 2.8 milliequivalents/l*
> *Precision:*
> > The mean of n determinations lies in the range, $\pm 0.02/n^{\frac{1}{2}}$ milliequivalents/l

B. Outline of method

A standard aliquot of seawater is mixed with an aliquot of standard acid. The pH before and after the treatment is used to calculate the total alkalinity and the carbonate alkalinity.

F. Experimental procedure

1. Standardize the pH meter using the phosphate buffer and measure the pH of the sample at room temperature. The original salinity and temperature of the sample (*in situ*) are also required for the calculation of carbonate alkalinity.

2. Pipette 25.00 ml of standard 0.01N HCl into a 200-ml wide mouth, screw-capped bottle and add a 100 ml aliquot of the seawater sample; stopper and mix.

3. Standardize the pH meter with the phthalate buffer at room temperature (20–25°C), rinse the electrodes with distilled water, and measure the pH of the acidified sample.

Calculation

1. Determine a_H, the hydrogen ion activity, as 10^{-pH} where the pH is the value measured above following acidification.

2. Find the value of f, the empirical activity coefficient, from Table 3 using the pH determined above and the salinity of the sample.

3. Calculate total alkalinity as (Note a):

$$\text{Total alkalinity (meq/l)} = 2.500 - 1250 \frac{a_H}{f} \text{ (see Note b)}$$

4. Calculate the initial pH of the seawater sample from the expression:

$$\text{pH (at } in\ situ \text{ temp. } t_2) = pH_{t_1} + 0.0114\,(t_1 - t_2)$$

where pH_{t_1} is the pH at laboratory temperature, t_1 (Note c).

5. Determine the carbonate alkalinity (milliequivalents per liter) as:

$$\text{carbonate alkalinity} = \text{total alkalinity} - A$$

where A is the milliequivalent correction for boric acid obtained in Table 5.

G. Total carbon dioxide content (all forms)

Total CO_2 content = carbonate alkalinity $\times F_T$ (millimoles/l) where F_T is found in Table 6.

CBMSA–F

TABLE 3. Factors for total alkalinity measurement. Factor f in the equation:

$$\text{total alkalinity} = 2.500 - 1250\,\frac{a_H}{f}$$

is found as a function of chlorinity or salinity. (Adapted from Strickland and Parsons, *Bull. Fish. Res. Bd. Canada*, **167**, 2nd ed., 1972.)

pH range	Cl‰ = 2 S‰ = 3.5	4 7	6 11	8 14.5	10 18	12–18 21–33	20 36
	f	f	f	f	f	f	f
2.8–2.9	0.865	0.800	0.785	0.775	0.770	0.768	0.773
3.0–3.9	0.845	0.782	0.770	0.760	0.755	0.753	0.758
4.0	0.890	0.822	0.810	0.800	0.795	0.793	0.798

TABLE 4. Total alkalinity calculation. Total alkalinity (meq/l at 20°C) as a function of the final pH of the solution obtained from 100.0 ml of sample and 25.00 ml of 0.0100N hydrochloric acid. This Table may be used for samples of chlorinity between 12‰ and 18‰, or salinities between 22‰ and 33‰. (Expanded from Strickland and Parsons, *Bull Fish. Res. Bd. Canada*, **167**, 2nd ed., 1972).

pH	Total alkalinity	pH	Total alkalinity	pH	Total alkalinity
2.80	(0.00)	3.20	1.45	3.60	2.08
2.82	0.08	3.22	1.50	3.62	2.10
2.84	0.18	3.24	1.55	3.64	2.12
2.86	0.27	3.26	1.59	3.66	2.14
2.88	0.36	3.28	1.63	3.68	2.15
2.90	0.45	3.30	1.67	3.70	2.17
2.92	0.54	3.32	1.71	3.72	2.18
2.94	0.61	3.34	1.74	3.74	2.20
2.96	0.69	3.36	1.77	3.76	2.21
2.98	0.76	3.38	1.81	3.78	2.23
3.00	0.84	3.40	1.84	3.80	2.24
3.02	0.92	3.42	1.87	3.82	2.25
3.04	0.99	3.44	1.90	3.84	2.26
3.06	1.06	3.46	1.93	3.86	2.27
3.08	1.12	3.48	1.95	3.88	2.28
3.10	1.19	3.50	1.98	3.90	2.29
3.12	1.24	3.52	2.00	3.92	2.30
3.14	1.30	3.54	2.02	3.94	2.31
3.16	1.35	3.56	2.04	3.96	2.32
3.18	1.40	3.58	2.06	3.98	2.33
				4.00	2.34

TABLE 5. Conversion of total alkalinity to carbonate alkalinity. Quantity A in milliequivalents per liter to be subtracted from total alkalinity to give carbonate alkalinity where pH and salinity are *in situ* values and tabulated values are multiplied by 10^{-2} to give A.

$S = 10\%_{0}$

pH:	7.3	7.4	7.5	7.6	7.7	7.8	7.9	8.0	8.1	8.2	8.3	8.4	8.5	8.6	8.7	8.8
$T=0°C$	0	0	0	0	0	0	1	1	1	1	1	2	2	2	3	3
$T=2$	0	0	0	0	0	0	1	1	1	1	1	2	2	2	3	3
$T=4$	0	0	0	0	0	0	1	1	1	1	1	2	2	3	3	4
$T=6$	0	0	0	0	0	1	1	1	1	1	1	2	2	3	3	4
$T=8$	0	0	0	0	0	1	1	1	1	1	2	2	2	3	3	4
$T=10$	0	0	0	0	0	1	1	1	1	1	2	2	2	3	3	4
$T=12$	0	0	0	0	0	1	1	1	1	1	2	2	2	3	3	4
$T=14$	0	0	0	0	0	1	1	1	1	1	2	2	3	3	4	4
$T=16$	0	0	0	0	1	1	1	1	1	1	2	2	3	3	4	4
$T=18$	0	0	0	0	1	1	1	1	1	2	2	2	3	3	4	4
$T=20$	0	0	0	0	1	1	1	1	1	2	2	2	3	3	4	5
$T=22$	0	0	0	0	1	1	1	1	1	2	2	2	3	3	4	5
$T=24$	0	0	0	0	1	1	1	1	1	2	2	3	3	4	4	5
$T=26$	0	0	0	1	1	1	1	1	1	2	2	3	3	4	4	5
$T=28$	0	0	0	1	1	1	1	1	2	2	2	3	3	4	4	5
$T=30$	0	0	0	1	1	1	1	1	2	2	2	3	3	4	4	5

$S = 20\%_{0}$

pH:	7.3	7.4	7.5	7.6	7.7	7.8	7.9	8.0	8.1	8.2	8.3	8.4	8.5	8.6	8.7	8.8
$T=0°C$	0	0	1	1	1	1	1	2	2	3	3	4	5	6	7	8
$T=2$	0	1	1	1	1	1	2	2	2	3	4	4	5	6	7	9
$T=4$	0	1	1	1	1	1	2	2	3	3	4	5	5	6	8	9
$T=6$	0	1	1	1	1	1	2	2	3	3	4	5	6	7	8	9
$T=8$	0	1	1	1	1	1	2	2	3	3	4	5	6	7	8	10
$T=10$	1	1	1	1	1	2	2	2	3	4	4	5	6	7	9	10
$T=12$	1	1	1	1	1	2	2	2	3	4	4	5	6	8	9	10
$T=14$	1	1	1	1	1	2	2	3	3	4	5	6	7	8	9	10
$T=16$	1	1	1	1	1	2	2	3	3	4	5	6	7	8	9	11
$T=18$	1	1	1	1	1	2	2	3	3	4	5	6	7	8	10	11
$T=20$	1	1	1	1	2	2	2	3	4	4	5	6	7	9	10	11
$T=22$	1	1	1	1	2	2	2	3	4	4	5	6	8	9	10	11
$T=24$	1	1	1	1	2	2	3	3	4	5	6	7	8	9	10	12
$T=26$	1	1	1	1	2	2	3	3	4	5	6	7	8	9	11	12
$T=28$	1	1	1	1	2	2	3	3	4	5	6	7	8	9	11	12
$T=30$	1	1	1	1	2	2	3	3	4	5	6	7	8	10	11	12

$S = 25\%_{0}$

pH:	7.3	7.4	7.5	7.6	7.7	7.8	7.9	8.0	8.1	8.2	8.3	8.4	8.5	8.6	8.7	8.8
$T=0°C$	1	1	1	1	1	2	2	3	3	4	5	6	7	8	10	11
$T=2$	1	1	1	1	1	2	2	3	3	4	5	6	7	9	10	12
$T=4$	1	1	1	1	2	2	2	3	4	4	5	6	8	9	10	12
$T=6$	1	1	1	1	2	2	2	3	4	5	5	7	8	9	11	12

146   *Gases in Seawater*

TABLE 5 (*cont.*)

$T=8$	1	1	1	1	2	2	3	3	4	5	6	7	8	10	11	13
$T=10$	1	1	1	1	2	2	3	3	4	5	6	7	8	10	12	13
$T=12$	1	1	1	1	2	2	3	3	4	5	6	7	9	10	12	14
$T=14$	1	1	1	2	2	2	3	4	4	5	6	8	9	11	12	14
$T=16$	1	1	1	2	2	2	3	4	5	6	7	8	9	11	13	14
$T=18$	1	1	1	2	2	3	3	4	5	6	7	8	10	11	13	15
$T=20$	1	1	1	2	2	3	3	4	5	6	7	8	10	12	13	15
$T=22$	1	1	1	2	2	3	3	4	5	6	7	9	10	12	14	15
$T=24$	1	1	2	2	2	3	4	4	5	6	8	9	11	12	14	16
$T=26$	1	1	2	2	2	3	4	4	5	7	8	9	11	12	14	16
$T=28$	1	1	2	2	3	3	4	5	6	7	8	9	11	13	14	16
$T=30$	1	1	2	2	3	3	4	5	6	7	8	10	11	13	15	16

$S=30‰$

pH:	7.3	7.4	7.5	7.6	7.7	7.8	7.9	8.0	8.1	8.2	8.3	8.4	8.5	8.6	8.7	8.8
$T=0°C$	1	1	1	1	2	2	3	3	4	5	6	8	9	11	12	14
$T=2$	1	1	1	2	2	2	3	4	4	5	7	8	9	11	13	15
$T=4$	1	1	1	2	2	2	3	4	5	6	7	8	10	12	13	15
$T=6$	1	1	1	2	2	3	3	4	5	6	7	9	10	12	14	16
$T=8$	1	1	1	2	2	3	3	4	5	6	7	9	11	12	14	16
$T=10$	1	1	2	2	2	3	4	4	5	6	8	9	11	13	15	17
$T=12$	1	1	2	2	2	3	4	5	6	7	8	10	11	13	15	17
$T=14$	1	1	2	2	3	3	4	5	6	7	8	10	12	14	16	18
$T=16$	1	1	2	2	3	3	4	5	6	7	9	10	12	14	16	18
$T=18$	1	1	2	2	3	3	4	5	6	8	9	11	12	14	17	19
$T=20$	1	2	2	2	3	4	4	5	6	8	9	11	13	15	17	19
$T=22$	1	2	2	2	3	4	5	6	7	8	10	11	13	15	17	19
$T=24$	1	2	2	3	3	4	5	6	7	8	10	12	14	16	18	20
$T=26$	1	2	2	3	3	4	5	6	7	9	10	12	14	16	18	20
$T=28$	1	2	2	3	3	4	5	6	7	9	10	12	14	16	18	20
$T=30$	1	2	2	3	3	4	5	6	8	9	11	13	14	17	19	21

$S=35‰$

pH:	7.3	7.4	7.5	7.6	7.7	7.8	7.9	8.0	8.1	8.2	8.3	8.4	8.5	8.6	8.7	8.8
$T=0°C$	1	1	1	2	2	3	4	4	5	7	8	9	11	13	15	18
$T=2$	1	1	2	2	2	3	4	5	6	7	8	10	12	14	16	18
$T=4$	1	1	2	2	3	3	4	5	6	7	9	10	12	14	17	19
$T=6$	1	1	2	2	3	3	4	5	6	8	9	11	13	15	17	20
$T=8$	1	1	2	2	3	4	4	5	6	8	9	11	13	15	18	20
$T=10$	1	2	2	2	3	4	5	6	7	8	10	12	14	16	18	21
$T=12$	1	2	2	3	3	4	5	6	7	9	10	12	14	16	19	21
$T=14$	1	2	2	3	3	4	5	6	7	9	11	13	15	17	19	22
$T=16$	1	2	2	3	3	4	5	6	8	9	11	13	15	17	20	22
$T=18$	1	2	2	3	4	4	5	7	8	10	11	13	16	18	20	23
$T=20$	2	2	2	3	4	5	6	7	8	10	12	14	16	18	21	23
$T=22$	2	2	3	3	4	5	6	7	8	10	12	14	16	19	21	24
$T=24$	2	2	3	3	4	5	6	7	9	10	12	15	17	19	22	24
$T=26$	2	2	3	3	4	5	6	7	9	11	13	15	17	20	22	24
$T=28$	2	2	3	3	4	5	6	8	9	11	13	15	18	20	22	25
$T=30$	2	2	3	4	4	5	7	8	10	11	13	16	18	20	23	25

TABLE 6. The factor (F_T) giving the total carbon dioxide content of seawater ($CO_2 + HCO_3^- + CO_3^=$) when multiplied by the carbonate alkalinity. Calculations are based on the values of K'_1 and K'_2 derived by Mehrbach *et al.* (Limnol. Oceanogr., **18**, 897, 1973).

$S = 5$

pH:	7.3	7.4	7.5	7.6	7.7	7.8	7.9	8.0	8.1	8.2	8.3	8.4	8.5	8.6	8.7	8.8
$T = 0°C$	1.15	1.12	1.09	1.07	1.06	1.05	1.04	1.03	1.02	1.02	1.01	1.01	1.00	1.00	1.00	0.99
$T = 2$	1.14	1.11	1.09	1.07	1.06	1.04	1.03	1.03	1.02	1.01	1.01	1.01	1.00	1.00	0.99	0.99
$T = 4$	1.13	1.11	1.08	1.07	1.05	1.04	1.03	1.02	1.02	1.01	1.01	1.00	1.00	1.00	0.99	0.99
$T = 6$	1.13	1.10	1.08	1.06	1.05	1.04	1.03	1.02	1.02	1.01	1.01	1.00	1.00	0.99	0.99	0.98
$T = 8$	1.12	1.10	1.08	1.06	1.05	1.04	1.03	1.02	1.01	1.01	1.00	1.00	0.99	0.99	0.98	0.98
$T = 10$	1.12	1.09	1.07	1.06	1.04	1.03	1.02	1.02	1.01	1.01	1.00	1.00	0.99	0.99	0.98	0.97
$T = 12$	1.11	1.09	1.07	1.05	1.04	1.03	1.02	1.02	1.01	1.00	1.00	0.99	0.99	0.98	0.97	0.96
$T = 14$	1.11	1.08	1.07	1.05	1.04	1.03	1.02	1.01	1.01	1.00	1.00	0.99	0.99	0.98	0.97	0.96
$T = 16$	1.10	1.08	1.06	1.05	1.04	1.03	1.02	1.01	1.00	1.00	0.99	0.98	0.98	0.97	0.96	0.95
$T = 18$	1.10	1.08	1.06	1.05	1.03	1.02	1.02	1.01	1.00	0.99	0.99	0.98	0.97	0.96	0.95	0.93
$T = 20$	1.09	1.07	1.06	1.04	1.03	1.02	1.01	1.00	1.00	0.99	0.98	0.97	0.96	0.95	0.94	0.92
$T = 22$	1.09	1.07	1.05	1.04	1.03	1.02	1.01	1.00	0.99	0.98	0.97	0.96	0.95	0.94	0.92	0.91
$T = 24$	1.09	1.07	1.05	1.04	1.02	1.01	1.00	1.00	0.99	0.98	0.97	0.95	0.94	0.93	0.91	0.89
$T = 26$	1.08	1.06	1.05	1.03	1.02	1.01	1.00	0.99	0.98	0.97	0.96	0.94	0.93	0.91	0.89	0.87
$T = 28$	1.08	1.06	1.04	1.03	1.02	1.01	1.00	0.98	0.97	0.96	0.95	0.93	0.92	0.90	0.88	0.85
$T = 30$	1.08	1.06	1.04	1.03	1.01	1.00	0.99	0.98	0.97	0.95	0.94	0.92	0.90	0.88	0.86	0.83

$S = 10$

pH:	7.3	7.4	7.5	7.6	7.7	7.8	7.9	8.0	8.1	8.2	8.3	8.4	8.5	8.6	8.7	8.8
$T = 0°C$	1.12	1.10	1.08	1.06	1.05	1.03	1.03	1.02	1.01	1.01	1.00	0.99	0.99	0.98	0.97	0.97
$T = 2$	1.12	1.09	1.07	1.06	1.04	1.03	1.02	1.02	1.01	1.00	1.00	0.99	0.98	0.98	0.97	0.96
$T = 4$	1.11	1.09	1.07	1.05	1.04	1.03	1.02	1.01	1.01	1.00	0.99	0.99	0.98	0.97	0.96	0.95
$T = 6$	1.10	1.08	1.06	1.05	1.04	1.03	1.02	1.01	1.00	1.00	0.99	0.98	0.98	0.97	0.96	0.95
$T = 8$	1.10	1.08	1.06	1.05	1.03	1.02	1.02	1.01	1.00	0.99	0.99	0.98	0.97	0.96	0.95	0.94
$T = 10$	1.09	1.07	1.06	1.04	1.03	1.02	1.01	1.01	1.00	0.99	0.98	0.98	0.97	0.96	0.94	0.93
$T = 12$	1.09	1.07	1.05	1.04	1.03	1.02	1.01	1.00	1.00	0.99	0.98	0.97	0.96	0.95	0.94	0.92
$T = 14$	1.09	1.07	1.05	1.04	1.03	1.02	1.01	1.00	0.99	0.98	0.97	0.96	0.95	0.94	0.93	0.91
$T = 16$	1.08	1.06	1.05	1.04	1.02	1.01	1.01	1.00	0.99	0.98	0.97	0.96	0.95	0.93	0.92	0.90
$T = 18$	1.08	1.06	1.05	1.03	1.02	1.01	1.00	0.99	0.98	0.97	0.96	0.95	0.94	0.92	0.91	0.89
$T = 20$	1.08	1.06	1.04	1.03	1.02	1.01	1.00	0.99	0.98	0.97	0.96	0.95	0.93	0.91	0.89	0.87
$T = 22$	1.07	1.06	1.04	1.03	1.02	1.01	1.00	0.99	0.98	0.96	0.95	0.94	0.92	0.90	0.88	0.86
$T = 24$	1.07	1.05	1.04	1.02	1.01	1.00	0.99	0.98	0.97	0.96	0.94	0.93	0.91	0.89	0.87	0.85
$T = 26$	1.07	1.05	1.03	1.02	1.01	1.00	0.99	0.98	0.97	0.95	0.94	0.92	0.90	0.88	0.86	0.83
$T = 28$	1.06	1.05	1.03	1.02	1.01	1.00	0.98	0.97	0.96	0.95	0.93	0.91	0.89	0.87	0.85	0.82
$T = 30$	1.06	1.04	1.03	1.02	1.00	0.99	0.98	0.97	0.95	0.94	0.92	0.90	0.88	0.86	0.84	0.81

$S = 20$

pH:	7.3	7.4	7.5	7.6	7.7	7.8	7.9	8.0	8.1	8.2	8.3	8.4	8.5	8.6	8.7	8.8
$T = 0°C$	1.09	1.07	1.06	1.04	1.03	1.02	1.01	1.00	0.99	0.98	0.97	0.96	0.95	0.93	0.92	0.90
$T = 2$	1.09	1.07	1.05	1.04	1.03	1.02	1.01	1.00	0.99	0.98	0.97	0.96	0.94	0.93	0.91	0.89
$T = 4$	1.08	1.06	1.05	1.03	1.02	1.01	1.00	0.99	0.98	0.97	0.96	0.95	0.94	0.92	0.90	0.88
$T = 6$	1.08	1.06	1.04	1.03	1.02	1.01	1.00	0.99	0.98	0.97	0.96	0.95	0.93	0.92	0.90	0.88
$T = 8$	1.07	1.06	1.04	1.03	1.02	1.01	1.00	0.99	0.98	0.97	0.96	0.94	0.93	0.91	0.89	0.87
$T = 10$	1.07	1.05	1.04	1.03	1.01	1.00	0.99	0.99	0.97	0.96	0.95	0.94	0.92	0.90	0.88	0.86
$T = 12$	1.07	1.05	1.04	1.02	1.01	1.00	0.99	0.98	0.97	0.96	0.95	0.93	0.91	0.90	0.87	0.85
$T = 14$	1.06	1.05	1.03	1.02	1.01	1.00	0.99	0.98	0.97	0.96	0.94	0.93	0.91	0.89	0.87	0.84
$T = 16$	1.06	1.04	1.03	1.02	1.01	1.00	0.99	0.98	0.96	0.95	0.94	0.92	0.90	0.88	0.86	0.83
$T = 18$	1.06	1.04	1.03	1.02	1.00	0.99	0.98	0.97	0.96	0.95	0.93	0.91	0.89	0.87	0.85	0.82
$T = 20$	1.06	1.04	1.03	1.01	1.00	0.99	0.98	0.97	0.96	0.94	0.92	0.91	0.89	0.86	0.84	0.81
$T = 22$	1.05	1.04	1.02	1.01	1.00	0.99	0.98	0.96	0.95	0.94	0.92	0.90	0.88	0.86	0.83	0.80
$T = 24$	1.05	1.03	1.02	1.01	1.00	0.99	0.97	0.96	0.95	0.93	0.91	0.89	0.87	0.85	0.82	0.79
$T = 26$	1.05	1.03	1.02	1.01	0.99	0.98	0.97	0.96	0.94	0.93	0.91	0.89	0.86	0.84	0.81	0.78
$T = 28$	1.05	1.03	1.02	1.00	0.99	0.98	0.97	0.95	0.94	0.92	0.90	0.88	0.86	0.83	0.80	0.78
$T = 30$	1.04	1.03	1.01	1.00	0.99	0.98	0.96	0.95	0.93	0.92	0.90	0.88	0.85	0.83	0.80	0.77

TABLE 6. (*cont.*)

S = 30

pH:	7.3	7.4	7.5	7.6	7.7	7.8	7.9	8.0	8.1	8.2	8.3	8.4	8.5	8.6	8.7	8.8
$T=0°C$	1.08	1.06	1.04	1.03	1.02	1.00	0.99	0.98	0.97	9.96	0.95	0.93	0.91	0.89	0.87	0.85
$T=2$	1.07	1.05	1.04	1.02	1.01	1.00	0.99	0.98	0.97	0.96	0.94	0.93	0.91	0.89	0.87	0.84
$T=4$	1.07	1.05	1.03	1.02	1.01	1.00	0.99	0.98	0.97	0.95	0.94	0.92	0.90	0.88	0.86	0.84
$T=6$	1.06	1.05	1.03	1.02	1.01	1.00	0.99	0.97	0.96	0.95	0.93	0.92	0.90	0.88	0.85	0.83
$T=8$	1.06	1.04	1.03	1.02	1.01	0.99	0.98	0.97	0.96	0.95	0.93	0.91	0.89	0.87	0.85	0.82
$T=10$	1.06	1.04	1.03	1.01	1.00	0.99	0.98	0.97	0.96	0.94	0.93	0.91	0.89	0.86	0.84	0.81
$T=12$	1.05	1.04	1.02	1.01	1.00	0.99	0.98	0.97	0.95	0.94	0.92	0.90	0.88	0.86	0.83	0.80
$T=14$	1.05	1.03	1.02	1.01	1.00	0.99	0.97	0.96	0.95	0.93	0.92	0.90	0.87	0.85	0.82	0.80
$T=16$	1.05	1.03	1.02	1.01	0.99	0.98	0.97	0.96	0.94	0.93	0.91	0.89	0.87	0.84	0.82	0.79
$T=18$	1.04	1.03	1.02	1.00	0.99	0.98	0.97	0.95	0.94	0.92	0.90	0.88	0.86	0.83	0.81	0.78
$T=20$	1.04	1.03	1.01	1.00	0.99	0.98	0.96	0.95	0.94	0.92	0.90	0.88	0.85	0.83	0.80	0.77
$T=22$	1.04	1.02	1.01	1.00	0.99	0.97	0.96	0.95	0.93	0.91	0.89	0.87	0.85	0.82	0.79	0.76
$T=24$	1.04	1.02	1.01	1.00	0.98	0.97	0.96	0.94	0.93	0.91	0.89	0.86	0.84	0.81	0.78	0.76
$T=26$	1.04	1.02	1.01	0.99	0.98	0.97	0.95	0.94	0.92	0.90	0.88	0.86	0.83	0.81	0.78	0.75
$T=28$	1.03	1.02	1.01	0.99	0.98	0.97	0.95	0.94	0.92	0.90	0.88	0.85	0.83	0.80	0.77	0.74
$T=30$	1.03	1.02	1.00	0.99	0.98	0.96	0.95	0.93	0.91	0.89	0.87	0.85	0.82	0.79	0.76	0.73

S = 35

pH:	7.3	7.4	7.5	7.6	7.7	7.8	7.9	8.0	8.1	8.2	8.3	8.4	8.5	8.6	8.7	8.8
$T=0°C$	1.07	1.05	1.04	1.02	1.01	1.00	0.99	0.98	0.96	0.95	0.94	0.92	0.90	0.88	0.86	0.83
$T=2$	1.07	1.05	1.03	1.02	1.01	1.00	0.99	0.97	0.96	0.95	0.93	0.92	0.90	0.87	0.85	0.82
$T=4$	1.06	1.04	1.03	1.02	1.01	0.99	0.98	0.97	0.96	0.94	0.93	0.91	0.89	0.87	0.84	0.82
$T=6$	1.06	1.04	1.03	1.01	1.00	0.99	0.98	0.97	0.96	0.94	0.92	0.91	0.89	0.86	0.84	0.81
$T=8$	1.05	1.04	1.02	1.01	1.00	0.99	0.98	0.97	0.95	0.94	0.92	0.90	0.88	0.86	0.83	0.80
$T=10$	1.05	1.04	1.02	1.01	1.00	0.99	0.97	0.96	0.95	0.93	0.92	0.90	0.87	0.85	0.82	0.80
$T=12$	1.05	1.03	1.02	1.01	0.99	0.98	0.97	0.96	0.94	0.93	0.91	0.89	0.87	0.84	0.82	0.79
T = 14	1.04	1.03	1.02	1.00	0.99	0.98	0.97	0.95	0.94	0.92	0.90	0.88	0.86	0.84	0.81	0.78
$T=16$	1.04	1.03	1.01	1.00	0.99	0.98	0.96	0.95	0.94	0.92	0.90	0.88	0.85	0.83	0.80	0.77
$T=18$	1.04	1.02	1.01	1.00	0.99	0.97	0.96	0.95	0.93	0.91	0.89	0.87	0.85	0.82	0.79	0.76
$T=20$	1.04	1.02	1.01	1.00	0.98	0.97	0.96	0.94	0.93	0.91	0.89	0.86	0.84	0.81	0.78	0.75
$T=22$	1.03	1.02	1.01	0.99	0.98	0.97	0.95	0.94	0.92	0.90	0.88	0.86	0.83	0.80	0.78	0.75
$T=24$	1.03	1.02	1.00	0.99	0.98	0.96	0.95	0.93	0.92	0.90	0.87	0.85	0.82	0.80	0.77	0.74
$T=26$	1.03	1.02	1.00	0.99	0.98	0.96	0.95	0.93	0.91	0.89	0.87	0.84	0.82	0.79	0.76	0.73
$T=28$	1.03	1.01	1.00	0.99	0.97	0.96	0.94	0.93	0.91	0.89	0.86	0.84	0.81	0.78	0.75	0.72
$T=30$	1.03	1.01	1.00	0.98	0.97	0.96	0.94	0.92	0.90	0.88	0.86	0.83	0.81	0.78	0.75	0.72

Notes

(a) If the pH is greater than 4.0 after the addition of acid, add a further 5.0 ml of 0.01N acid and calculate total alkalinity as:

$$\text{total alkalinity (meq)} = 3.00 - 1300 \frac{a_H}{f}$$

(b) For salinities between 22 and 33‰, the total alkalinity can be found from Table 4 without introducing appreciable error.

(c) The temperature correction equation comes from Gieskes (*Limnol. Oceanogr.*, **14**: 679, 1969) and is independent of salinity within the precision of the method.

7.3. Determination of Sulfide

Introduction

The determination of dissolved sulfide (H_2S, HS^-, S^{2-}) in seawater has usually been accomplished by the methylene blue method (e.g., Kato *et al.*, *Technol. Rept.*, *Tohoku Univ.*, **19**: 85, 1954). Recently, a method which has several advantages over this procedure has been developed by Cline (*Limnol. Oceanogr.*, **14**: 454, 1969).

Method

A. Capabilities

    Range: 1.5 to 1000 µg-at/l
    1. Precision at the 3 µg-at/l level:
    The correct value lies in the range, mean of *n* determinations $\pm 0.22/n^{\frac{1}{2}}$ µg-at/l
    2. `Precision at the 20 µg-at/l level:
    The correct value lies in the range, mean of *n* determinations $\pm 3.0/n^{\frac{1}{2}}$ µg-at/l

B. Outline of method

A seawater sample is treated with a mixed solution of N,N-dimethyl-p-phenylenediamine sulfate and ferric chloride in acidic medium. Methylene blue is formed over time and is measured spectrophotometrically.

C. Special apparatus and equipment

50-ml capacity, stoppered, graduated glass cylinder
125-ml glass-stoppered reagent bottle
Automatic pipette with 4-ml stop position

D. Sampling procedure and storage

Since hydrogen sulfide is volatile and any sulfide is rapidly oxidized by dissolved oxygen in water or upon exposure to air, precautions regarding sample collection are required as described in Method 7.1. The samples for sulfide should be drawn into 125-ml tightly stoppered reagent bottles using a long rubber spigot. Analyses should be commenced within an hour of taking the samples, which should meantime be stored in a cool, dark place.

E. Special reagents

Mixed diamine reagent solution

Dissolve the amounts of N,N-dimethyl-p-phenylenediamine sulfate (e.g., Eastman Kodak No. 1333) and analytical reagent quality ferric chloride ($FeCl_3 \cdot 6H_2O$) shown in the following table in 500 ml of analytical quality, cooled, 6N hydrochloric acid. The mixed reagent solution will be stable for several months in dark bottles under refrigeration, except for the most dilute reagent.

TABLE 7. Suggested reagent concentrations and dilution factors to be used in the determination of sulfide-sulfur in the stated concentration range.

Sulfide concen. (μmoles/l)	(A) Diamine concen. (g/500 ml)	(B) Ferric concen. (g/500 ml)	Ratio of reagents A and B to be mixed	Light path length (cm)
1–3	0.5	0.75	1:1	10
3–40	2.0	3.0	1:1	1
40–250	8.0	12.0	2:25	1
250–1000	20.0	30.0	1:50	1

F. Experimental procedure

1. Add 50 ml of sample to a dry 50-ml graduated cylinder fitted with a stopper. Do not allow the sample to be aerated with bubbles during this addition.

2. Without delay, add 4 ml of the appropriate mixed diamine reagent

solution from an automatic pipette, stopper and invert the cylinder twice (Notes a, b).

3. Measure the extinction of the solution against distilled water at 670 nm in the appropriate cuvette after 20 min (Note c).

4. After color development, dilute the solution, if the extinction is greater than 1.0, using appropriate capacity volumetric glassware (Note d).

5. Calculate the sulfide concentration in microgram-atoms of sulfur per liter from the expression:

$$\mu\text{g-at S/l} = F \times (E_s - E_b)$$

where E_s is the extinction of a sample, E_b is a blank extinction as described in Section G, and F is the factor obtained as described in Section H, below.

G. Determination of blank

The amount of sulfide in ordinary, oxygenated, surface seawater may be considered negligible. Use filtered, surface seawater and carry out the method exactly as described in Section F, steps 1–4 inclusive. The value of the blank extinction will depend on reagent concentration and purity, sample turbidity, and cell to cell differences.

H. Calibration

1. Oxygen-free distilled water

Prepare oxygen-free water by purging distilled water with nitrogen from a cylinder for at least 30 min. If cylinder nitrogen is not available, use boiled and cooled distilled water.

2. Standard sulfide solution (prepare fresh before use)

Add 1.2 g of analytical grade sodium sulfide ($Na_2S \cdot 9H_2O$) into 100 ml of the oxygen-free distilled water (Note e). Dissolve the sulfide under a nitrogen atmosphere. The concentration of the standard can be calculated from the amount of sulfide added.

$$1 \text{ ml} \equiv 50 \ \mu\text{g-at sulfur}$$

However, a more accurate estimation of sulfide should be examined by standardized iodometry as described below.

3. 0.1N sodium thiosulfate solution

Dissolve 25 g of analytical grade sodium thiosulfate ($Na_2S_2O_3 \cdot 5H_2O$)

and 0.1 g of sodium carbonate (Na_2CO_3) in 1 l of boiled and cooled distilled water.

4. 0.1N iodine solution

Dissolve 6.4 g of analytical reagent quality iodine (I_2) in 25 ml of distilled water containing *ca.* 40 g of potassium iodide (KI). When the iodine has dissolved, dilute the solution to 500 ml with distilled water. Store the solution in the dark and refrigerated.

5. Standardization of sulfide

Pipette 10 ml of the sulfide solution into a beaker containing 250 ml of oxygen-free distilled water. Add 20 ml of 0.1N iodine and 25 ml of 0.1N HCl (the acid need not be standardized). Titrate excess iodine with 0.1N thiosulfate, using starch as an indicator. Calculate the milliliters of 0.1N iodine required to react with the sulfide.

One ml of 0.1N iodine is equivalent 50 μg-atoms of sulfide-sulfur. Calculate the volume of sulfide solution equivalent to 2, 20, 200, and 500 μg-atoms/l, and dilute this volume to 500 ml with the oxygen-free distilled water in order to prepare standards for concentration ranges given in Table 7.

6. Procedure

Pour 50 ml of standard sulfide solution with minimum aeration into a dry 50-ml measuring cylinder. Add 4 ml of mixed diamine reagent solution from an automatic pipette. Mix by gently inverting once or twice and continue the method as described in Section F above. Calculate the factor F (for 10- or 1-cm cuvettes) from the expression:

$$F = \frac{c}{E_s - E_b}$$

where E_s is the mean extinction of at least three standards, E_b is the mean extinction of at least two blanks, and c is the concentration of sulfide-sulfur (in μg-at/l) which was corrected by the calculation as described in paragraph 5 above.

Notes

(a) Stable color development is given over a wide temperature range (> 5–$30°C$), if a single solution in which a mixture of acidic diamine and ferric chloride are used in order to avoid volatilization of H_2S (see Cline, *loc. cit.*).

(b) After transferring solutions to the 50-ml cylinders, even with as little oxygenation as possible, sulfide oxidation commences and can be detected even after 5 to 10 min. The analysis should not be delayed more than a few minutes at this point.

(c) The color development should be complete within 10 min, and solutions are therefore stable for days although the extinction may tend to increase slightly.

(d) The solutions can be diluted with distilled water after color development to give a strictly proportional decrease of extinction. This method is without salt effect over the salinity range of 0–40‰. However, aqueous methylene blue solutions do not conform strictly to Beer's law at extinctions greater than 1.0.

(e) Lumps of sodium sulfide should be thoroughly washed with oxygen-free water to remove any sodium sulfite and wiped with a cellulose tissue immediately before preparation of the standard solution. Precaution should be taken throughout the preparation of the standard to protect the sodium sulfide from contact with oxygen.

Counting, Media and Preservatives

8.1. Collection and Enumeration of Organisms

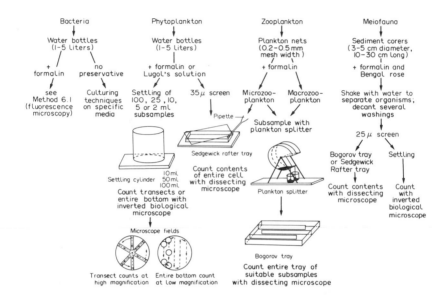

Confidence Limits of Counting for 95% Probability: (Note a)

The confidence limits of a single count, assuming a Poisson distribution of counts, can be given by the graph shown in Fig. 7 for counts of 0 to 100 organisms. For replicate counts (n), in which the mean and variance are calculated, the confidence limits can be calculated from t-tables for a given number of degrees of freedom (n-1).

Notes

(a) The confidence limits given assume a Poisson distribution of the number of organisms counted in the counting chamber. A Chi-square test for the correctness of this assumption can be made on replicate counts and alternative confidence limits applied if necessary. Generally, however, subsampling and counting can be carried out in order to avoid obvious clumping of organisms. References to non-random statistics of counting are given by Holmes and Widrig (*J. Cons. Explor. Mer*, **22**: 21, 1956) and Kutkuhn (*Limnol. Oceanogr.*, **3**: 69, 1958).

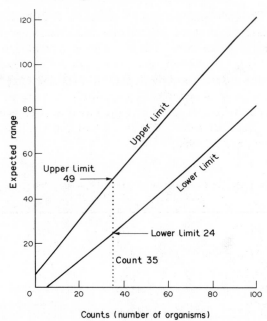

FIGURE 7. The 0 95 confidence limits for the expectation of a Poisson variable. For the confidence limits of a single count, find the count on the horizontal axis and draw a vertical line through it as shown in the example for a count of 35. (Redrawn from Lund *et al.*, *Hydrobiologia*, **11**: 143, 1958).

Additional References:

Schlieper, C. 1972. *Research methods in marine biology*. Sidgwick & Jackson, London. 356 pp.
Sournia, A. (ed.). 1978. *Monographs on oceanographic methodology*. No. 6. Phytoplankton manual. Unesco Press, Paris. 337 pp.
UNESCO. 1968. *Monographs on oceanographic methodology*. No. 2. Zooplankton sampling. Unesco Press, Paris. 174 pp.

8.2. Artificial Seawater Media

Introduction

Many different salt and nutrient concentrations have been used for the cultivation of marine algae. In general, these can be divided into media which are based on natural seawater to which nutrients have been added, and those based upon an artificial seawater composed of distilled water with additions of the major ions. The following media (Procedure, 1 & 2) both use the same nutrient additions (HES), but one is based on artificial seawater (AS) (Table 8) and one on natural seawater (NW). The recipe is taken from Harrison *et al. (J. Phycol.*, **16**: 28, 1980) and has been used successfully to culture approximately 100 different species and strains of phytoplankton, representing eleven classes of algae. A discussion of these media is given in the reference above; similar media, upon which the recipes given here are based in part, will be found in Provasoli *et al. (Arch. Mikrobiol.*, **25**: 392, 1960) and Guillard and Ryther (*Can. J. Microbiol.*, **8**: 229, 1962).

TABLE 8. Composition of Artificial Seawater (AS)
$S = 30.5\%_0$; sp. gr. = 1.021 at 20°C.

Solution $I^{(b)}$	MW	$g/kg^{(a)}$
NaCl	58.44	20.758
Na_2SO_4	142.04	3.477
KCl	74.56	0.587
$NaHCO_3$	84.00	0.170
KBr	119.01	0.0845
H_3BO_3	61.83	0.0225
NaF	41.99	0.0027
Solution $II^{(b)}$		
$MgCl_2 \cdot 6H_2O$	203.33	9.395
$CaCl_2 \cdot 2H_2O$	147.33	1.316
$SrCl_2 \cdot 6H_2O$	266.64	0.0214

(a) In practice, it is convenient to make up the final volume of the salt solution and nutrients to 1 l with distilled water, which results in slightly more than 1 kg due to the change in density.

(b) Solutions I and II are made up separately, autoclaved at 15 psi for 30 min and combined to give a total volume of 1 l.

Procedure

1. Artificial seawater with nutrients (HESAW)

Add 10 ml of solution III and 1.0 ml of solution IV (see Table 9) to 1 l of artificial seawater (solutions I and II). Mix and dispense aseptically. (Note a)

2. Natural seawater (Note b) with nutrients (HESNW)

Filter several liters of seawater through a HA Millipore filter and allow to age in a dark, cold room for several months. Autoclave a 1-l sample for 30 min at 15 psi, allow to cool (Note c) and add 10 ml of solution III and 1.0 ml of solution IV (see Table 9). Mix and dispense aseptically. (Note a)

TABLE 9. Composition in g/l of two nutrient enriched
solutions (HES).

Solution III (Nutrients and trace metals)[a]	g/l	Final conc. of a 10^{-2} dilution (μM)
NaNO$_3$	4.667	549.09
Na$_2$SiO$_3 \cdot$9H$_2$O[b]	3.000	105.60
Na$_2$glyceroPO$_4$	0.667	21.79
Na$_2$EDTA[c]	0.553	14.86
H$_3$BO$_3$	0.380	61.46
Fe(NH$_4$)$_2$(SO$_4$)$_2 \cdot$6H$_2$O(c)	0.234	5.97
FeCl$_3 \cdot$6H$_2$O(c)	0.016	0.59
MnSO$_4 \cdot$4H$_2$O	0.054	2.42
ZnSO$_4 \cdot$7H$_2$O	0.0073	0.25
CoSO$_4 \cdot$7H$_2$O	0.0016	0.06
Solution IV (Vitamins)[d]		
Thiamine	0.1	0.297
Vitamin B$_{12}$	0.002	0.0015
Biotin	0.001	0.0041

(a) Sterilize by autoclaving for 30 min at 15 psi. Store in a dark bottle in the refrigerator.

(b) The silicate must be neutralized before this solution is added to natural (NW) or artificial (AW) seawater. To 100 ml of silicate solution containing 30g/l, add 20 ml of 1N HCl and dilute immediately with the other additives.

(c) Na$_2$EDTA, FeCl$_3 \cdot$6H$_2$O, and Fe(NH$_4$)$_2$(SO$_4$)$_2 \cdot$6H$_2$O must be in solution before the remaining trace metals (Mn, Zn, Co) are added.

(d) Filter, sterilize, and store frozen.

Notes

(a) Final nutrient concentrations are given in μmol/l (μM) in the last column of Table 9. These concentrations are suitable for maintaining stock cultures. If nutrient solutions are required which have much lower concentrations (i.e., comparable to natural levels), the quantity of solutions III and IV added can be adjusted to one-tenth or one-hundredth the quantities suggested above.

(b) Because of the expense and time involved in making up artificial seawater, stock cultures are often more conveniently maintained on natural seawater (HESNW). Artificial seawater is particularly important, however, if a well defined medium is needed which does not contain the natural organic compounds (as well as some potential pollutants) which may be present in natural seawater samples.

(c) For the routine maintenance of stock cultures, it may be more convenient to mix the aged seawater and solutions III and IV, and then autoclave at 15 psi for 30 min. This results in the destruction of some vitamins and produces a small amount of precipitate. However, since the solution contains an excess in nutrient enrichment, the loss of some fraction of these compounds may not affect the growth of the algae. If this procedure is followed, the absolute concentration of nutrients will not be known.

8.2.1. Culture Collections

PARTIAL LIST OF ADDRESSES WHERE ALGAL CULTURES MAY BE OBTAINED (from: *Selected Papers in Phycology, Ed.* J.R. Rosowski and B.C. Parker, 1982, *Publ.* by Phycological Soc. Amer. Inc.; Lawrence, Kansas, U.S.A.)

Australia
Dr. S.W. Jeffrey
C.S.I.R.O. Marine Laboratories
Box 21, Cronulla, NSW 2230
Australia

Canada
Ms. J. Acreman
Department of Oceanography
University of British Columbia
Vancouver, B.C.
Canada V6T 1W5

Dr. D.V. Subba Rao
P.O. Box 1006
Bedford Institute of Oceanography
Dartmouth, N.S.
Canada B2Y 4A2

England
Dr. J.C. Green
Marine Biological Association of the United Kingdom
Citadel Hill
Plymouth
Devon PL1 2PB
England

France
   Mme. le Professeur P. Gayral
   Laboratoire d'Algologie Fondamentale et Appliquée
   Université de Caen
   39 Rue Desmoueux
   1400 Caen
   France

Japan
   Dr. Yuzaburo Ishida
   Dept. of Fisheries
   Faculty of Agriculture
   Kyoto University
   Kyoto
   Japan

People's Republic of China
   Dr. C.K. Tseng
   Institute of Oceanology
   Academia Sinica
   Qingdao
   Shangdong
   P.R.C.

Soviet Union
   Dr. L.A. Lanskaya
   Institute for the Biology of South Seas
   Academy of Sciences, UkSSR
   2 Nakhimov Ave.
   Sebastopol 335000
   U.S.S.R.

United States
   Dr. Greta Fryxell
   Dept. of Oceanography
   Texas A and M University
   College Station
   Texas
   U.S.A. 77843

   Dr. R.R.L. Guillard
   Bigelow Laboratory for Ocean Sciences
   McKown Point
   West Boothbay Harbour
   Maine
   U.S.A. 04575

Dr. J. Lewin
School of Oceanography
University of Washington
Seattle, Washington
U.S.A. 98195

8.3. Plankton Preservatives

A. *Phytoplankton*

I. Lugol's Solution
Prepare a stock solution of the following:

> 200 g KI
> 100 g I_2
> 2000 ml H_2O
> 190 ml glacial acetic acid

Add 10 drops of Lugol's solution to each 200 ml of phytoplankton sample.

B. *Zooplankton*

I. *For routine use with mixed samples:*
Dilute concentrated formalin (40% formaldehyde as purchased) to about 4 or 5% formalin ($= ca.$ 2% formaldehyde). This should be used in a ratio of 1 part plankton biomass to 9 parts preservative fluid. The calcareous skeletons of certain zooplankton will dissolve rapidly in unbuffered formalin; therefore, it may be desirable to use one of the preservatives discussed below.

II. *For long-term preservation of calcareous zooplankton (foraminifera, mollusca):*

  (a) *Hexamine-buffered formalin*
Prepare a stock solution of 40% formaldehyde supersaturated with hexamethylene tetramine ("hexamine"); this requires approximately 200 g of hexamine/liter of formaldehyde. For preservation, dilute the stock solution to 4 or 5% (see BI, above) and add to the plankton sample in a ratio of 9 parts preservative to 1 part plankton biomass.

  (b) *Borax-buffered formalin*
Add 30 g of sodium tetraborate ($Na_2B_4O_7 \cdot 10H_2O$ or borax) to 1 l of 40% formaldehyde. Dilute this stock solution to 4 or 5% (see BI, above) and use 9 parts of preservative to 1 part of plankton biomass.

Additional References:

Steedman, H.F. (ed.). 1976. *Monographs on oceanographic methodology*. No. 4. Zooplankton fixation and preservation. Unesco Press, Paris. 350 pp.

Terms and Equivalents

9.1. Oxygen Equivalents (Chemical and Biological)

(a) ml $O_2/l = 11.2 \times$ mg-at O_2/l

(b) mg $O_2/l = 16.0 \times$ mg-at O_2/l

(c) 1 ml of O_2 consumed $\equiv$ 5 g-cal assuming an RQ of approximately 1.0

(d) mg O_2 consumed per unit time $\times \frac{12}{32} \times$ RQ $\equiv$ mg C utilized per unit time

(e) ml O_2 (NTP) consumed per unit time $\times \frac{12}{22 \cdot 4} \times$ RQ

$$\equiv \text{mg C utilized per unit time}$$

where RQ in (c), (d) and (e) is the molecular respiratory quotient, $+ \triangle CO_2 / - \triangle O_2$, which may range in animals from 0.7 to 1.0 depending upon whether fats or carbohydrates are being utilized for energy, respectively.

(f) mg C photosynthesized per unit time $\equiv$

$$\frac{\text{mg } O_2 \text{ produced per unit time}}{PQ} \times \frac{32}{12}$$

where PQ is the molecular photosynthetic quotient, $+ \triangle O_2 / - \triangle CO_2$, which may range from *ca.* 1 to 2 but for phytoplankton is generally in the range 1 to 1.3 (Ryther, *Limnol. Oceanogr.*, **1**: 72, 1956).

9.2. Energy Equivalents

(a) 1 mg of food (dry weight) $\equiv$ 5.5 g-cal, assuming an average dry weight composition of 50% protein, 20% fat, 20% carbohydrate, 10% ash and caloric equivalents of 5.65 cal/mg for protein, 4.1 cal/mg for carbohydrate and 9.45 cal/mg for fat.

(b) Zooplankton

(i) cal/g dry wt $\equiv -227 + 152(\%$ carbon) for $r = 0.94$ and 21 d.f.

(ii) cal/g dry wt $\equiv 1351 + 106(\%$ carbon) $- 21.2(\%$ ash) for $r = 0.98$ and 16 d.f.

Data in (i) and (ii) from Platt *et al.* (*J. Fish. Res. Bd. Canada*, **26**: 2345,

1969) for total zooplankton collected over 1 year at 44°35′N, 64°02′W.

(c) Phytoplankton

   (i) cal/mg dry wt $\equiv 0.632 \times 0.086$(% carbon) for $r = 0.95$ and 9 d.f.

   (ii) cal/mg dry wt $\equiv = -0.555 + 0.113$(% carbon) $+ 0.054$(C:N ratio) for r = 0.96 and 9 d.f.

Data in (i) and (ii) from Platt and Irwin (*Limnol. Oceanogr.*, **18**: 306, 1973) for samples of natural phytoplankton collected during a spring bloom at 44°35′N, 64°02′W.

9.3. Carbon Equivalents

(a) Phytoplankton

   (i) For phytoplankton, other than diatoms:

$$\log C \equiv 0.866 \log V - 0.460$$

where C is the carbon per cell in picograms and V is the cell volume in cubic microns, for $n = 13$ and 95% confidence limits for the slope, 0.866 ± 0.080.

   (ii) For diatoms:

$$\log C \equiv 0.758 \log V - 0.422$$

where C is the carbon per cell in picograms and V is the cell volume in cubic microns, for $n = 96$ and 95% confidence limits for the slope, 0.758 ± 0.032.

Data in (i) and (ii) above from Strathmann (*Limnol. Oceanogr.*, **12**: 411, 1967).

(b) Zooplankton

   (i) For mixed samples containing more than 90% copepods:

$$\text{mg carbon} \equiv 0.49 \text{ mg dry weight} - 5.19$$

For $r = 0.99$, $n = 45$. Corrected equivalent given by Fulton (pers. comm.) from *Fish. Res. Bd. Canada* Tech. Rept. No. 313, 1972.

   (ii) For *Euphausia pacifica*:

$$\text{mg carbon} \equiv 0.42 \text{ mg dry weight}$$

For $n = 22$ and 95% confidence limits of 0.42 ± 0.03. From Lasker (*J. Fish. Res. Bd. Canada*, **23**: 1291, 1966).

(iii) Range of major chemical constituents of zooplankton:

Organism or group	Dry wt. as % wet wt.	Carbon	Nitrogen	Hydrogen	Phosphorus	Ash	Comment
		(all constituents expressed as a % of dry wt.)					
Copepods*	11.6–16.3	35–48	8.2–11.2		0.7–0.8		Sargasso Sea
Euphausiids &							plankton from
mysids*	14.5–18.0	35–43	9.4–10.5		1.4–1.6		Beers (1966)—
Chaetognaths*	6.0– 7.4	22–34	6.3– 9.4		0.5–0.7		range for each group
Fish/fish larvae*	11.9–16.0	33–42	8.3–10.7		0.9–1.8		
Polychaetes*	5.7–27.0	16–44	4.4–11.2		0.4–1.8		
Siphonophores*	0.3– 6.1	3–16	1.0– 4.4		<0.1–0.2		
Hydromedusae*	0.3–10.1	5–10	1.4– 6.2		0.1–0.4		
Pteropods*	22–32	21–25	2.7– 4.2		0.2–0.4		
Copepods+	10.2–15.8	32–42	4.7– 7.1		0.4–0.8	18–23	Continental
Euphausiids+	19.0–20.0	33–37	5.2– 7.1		0.9–1.2	19–22	shelf off New
Ctenophores+	4.7– 5.0	(6.4)	0.2– 1.1		0.1–0.2	70–75	York, from
Pteropods+	3.5–19.0	26–28	2.2– 5.0		0.3–0.6	24–64	Curl (1962)—
Tunicates+	4.0– 4.1	7–11	0.3– 1.5		0.1–0.3	71–77	range for each group
Copepods*	9.2–33.9	39–66	5.1–13.1	6.7–10.3		2–6	North Pacific
Amphipods*	18.4–36.6	26–48	4.4– 8.2	4.4– 7.6		10–37	from Omori
Euphausiids*	20.2–21.3	39–47	10.0–10.7	6.7– 7.6		8–9	(1969)—range
Chaetognaths*	11.6–14.1	44–48	10.7–11.1	7.2– 7.6		4–5	for each group
Pteropods*	25.0–36.4	17–29	1.5– 6.0	1.1– 3.8		29–43	

*Dried at 60 C. +Dried at 105 C.

Above table from Beers (*Limnol. Oceanogr.*, **11**: 520, 1966), Curl (*Rapp. Proc.-Verb. Cons. int. Explor. Mer*, **153**: 183, 1962), Omori (*Mar. Biol.*, **3**: 4, 1969) as compiled in Parsons, Takahasi and Hargrave, *Biological Oceanographic Processes*, Pergamon Press (Oxford), 1977.

9.4. Metric Units and Equivalents

Metric Prefixes:

atto (a)	= 0.000000000000000001	= 10^{-18}
femto (f)	= 0.000000000000001	= 10^{-15}
pico (p)	= 0.000000000001	= 10^{-12}
nano (n)	= 0.000000001	= 10^{-9}
micro (μ)	= 0.000001	= 10^{-6}
milli (m)	= 0·001	= 10^{-3}
centi (c)	= 0.01	= 10^{-2}
deci (d)	= 0.1	= 10^{-1}
deca (da)	= 10	= 10^{1}
hecto (h)	= 100	= 10^{2}
kilo (k)	= 1000	= 10^{3}
mega (M)	= 1000000	= 10^{6}
giga (G)	= 1000000000	= 10^{9}
tera (T)	= 1000000000000	= 10^{12}

Units of Length:

1 ångström (Å)	= 0.0001 micron
1 nanometer (nm)	= 10^{-9} meter
1 micron (μ)	= 0.001 millimeter (or 10^{-3} mm)
	1 μmeter
1 millimeter (mm)	= 1000 microns
	0.001 meter
1 centimeter (cm)	= 10 millimeters
	0.394 inch
1 decimeter (dm)	= 0.1 meter
1 meter (m)	= 100 centimeters
	3.28 feet
1 kilometer (km)	= 1000 meters
	3280 feet
	0.62 statute mile
	0.54 nautical mile
1 inch	= 2.54 centimeters
1 foot	= 0.3048 meter
1 yard	= 3 feet
	0.91 meter
1 fathom	= 6 feet
	1.83 meters
1 statute mile	= 5280 feet
	1.6 kilometers
	0.87 nautical mile
1 nautical mile	= 6076 feet
	1.85 kilometers
	1.15 statute miles

Units of Area:

1 square centimeter (cm^2)	= 100 square millimeters
	0.155 square inch
1 square meter (m^2)	= 10^4 square centimeters
	10.8 square feet
1 are (a)	= 100 square meters
1 square kilometer (km^2)	= 10^6 square meters
	247.1 acres
	0.386 square statute mile
	0.292 square nautical mile
1 hectare (ha)	= 10000 square meters

Units of Speed:

1 knot	= 1 nautical mile/hour
	1.15 statute miles/hour
	1.85 kilometers/hour
Velocity of sound in water of salinity = 35‰	= 1507 meters/second

Units of Volume:

1 milliliter (ml)	= 0.001 liter
	1 cm^3 (or 1 cc)
1 liter (l)	= 1000 cm^3
	1.06 liquid quarts
1 gallon	= 4.55 dm^3 = 4.55 liters
1 cubic meter (m^3)	= 1000 liters

Units of Mass:

1 milligram (mg)	= 0.001 gram
1 kilogram (kg)	1000 grams
	2.2 pounds
1 tonne (t)	= 1 metric ton
	10^6 grams
	1 Mg
1 pound	= 453.6 grams
1 long ton (UK)	= 1.016 Mg
1 short ton (US)	= 0.907 Mg

Units of Concentration:

M	$\equiv$ gram molecular weight per liter (or molar concentration)
μg-atoms/l	$\equiv$ mg-atoms/m^3
	μM
μg/l	$\equiv$ mg/m^3
ppm (parts per million)	$\equiv$ mg/l
ppb (parts per billion)	$\equiv$ μg/l
μg/l ÷ atomic wt	= μg-atoms/l
μg/l ÷ molecular wt	= μM = μmol/l

Terms and Equivalents

9.5. Sigma-T Values

Table 10. Sigma-T (σ_t) where density is given as a function of temperature (C°) and salinity (‰).

Note: To obtain the density of seawater:

$$\text{Density} = [(\sigma_t \times 10^{-3}) + 1] \text{ g/cm}^3.$$

S(‰)						T(°C)								
	0	1	2	3	4	5	6	7	8	9	10	11	12	13
0	−0.07	−0.01	0.03	0.05	0.06	0.05	0.03	−0.01	−0.06	−0.13	−0.21	−0.31	−0.41	−0.54
1	0.74	0.80	0.84	0.86	0.86	0.85	0.82	0.78	0.72	0.65	0.57	0.47	0.36	0.24
2	1.56	1.61	1.64	1.66	1.66	1.65	1.62	1.57	1.51	1.44	1.35	1.25	1.14	1.01
3	2.37	2.42	2.45	2.46	2.46	2.44	2.41	2.36	2.30	2.22	2.13	2.03	1.91	1.79
4	3.18	3.22	3.25	3.26	3.25	3.23	3.20	3.15	3.08	3.00	2.91	2.81	2.69	2.56
5	3.99	4.03	4.05	4.06	4.05	4.03	3.99	3.93	3.87	3.78	3.69	3.58	3.46	3.33
6	4.80	4.84	4.86	4.86	4.85	4.82	4.78	4.72	4.65	4.57	4.47	4.36	4.24	4.10
7	5.61	5.64	5.66	5.66	5.64	5.61	5.57	5.51	5.43	5.35	5.25	5.13	5.01	4.87
8	6.42	6.45	6.46	6.45	6.44	6.40	6.35	6.29	6.22	6.13	6.02	5.91	5.78	5.64
9	7.22	7.25	7.26	7.25	7.23	7.19	7.14	7.08	7.00	6.19	6.80	6.68	6.56	6.41
10	8.03	8.05	8.06	8.05	8.02	7.98	7.93	7.86	7.78	7.69	7.58	7.46	7.33	7.18
11	8.84	8.86	8.86	8.84	8.82	8.77	8.72	8.65	8.56	8.46	8.35	8.23	8.10	7.95
12	9.64	9.66	9.66	9.64	9.61	9.56	9.50	9.43	9.34	9.24	9.13	9.01	8.87	8.72
13	10.45	10.46	10.46	10.44	10.40	10.35	10.29	10.21	10.12	10.02	9.91	9.78	9.64	9.49
14	11.25	11.26	11.25	11.23	11.19	11.14	11.07	11.00	10.90	10.80	10.68	10.55	10.41	10.26
15	12.06	12.06	12.05	12.02	11.98	11.93	11.86	11.78	11.63	11.58	11.46	11.33	11.18	11.03
16	12.86	12.86	12.85	12.82	12.77	12.72	12.65	12.56	12.46	12.35	12.23	12.10	11.95	11.80
17	13.67	13.66	13.65	13.61	13.57	13.51	13.43	13.34	13.24	13.13	13.01	12.87	12.72	12.56
18	14.47	14.46	14.44	14.41	14.36	14.29	14.22	14.13	14.02	13.91	13.78	13.64	13.49	13.33
19	15.27	15.26	15.24	15.20	15.15	15.08	15.00	14.91	14.80	14.69	14.56	14.42	14.26	14.10
20	16.08	16.06	16.04	15.99	15.94	15.87	15.79	15.69	15.58	15.46	15.33	15.19	15.03	14.87
21	16.88	16.86	16.83	16.79	16.73	16.66	16.57	16.47	16.36	16.24	16.11	15.96	15.80	15.64
22	17.68	17.66	17.63	17.58	17.52	17.44	17.35	17.25	17.14	17.02	16.88	16.73	16.57	16.40
23	18.48	18.46	18.42	18.37	18.31	18.23	18.14	18.04	17.92	17.79	17.65	17.50	17.34	17.17
24	19.29	19.26	19.22	19.17	19.10	19.02	18.92	18.82	18.70	18.57	18.43	18.28	18.11	17.94
25	20.09	20.06	20.02	19.96	19.89	19.81	19.71	19.60	19.48	19.35	19.20	19.05	18.88	18.71
26	20.89	20.86	20.81	20.75	20.68	20.59	20.49	20.38	20.26	20.13	19.98	19.82	19.66	19.48
27	21.70	21.66	21.61	21.55	21.47	21.38	21.28	21.17	21.04	20.90	20.75	20.60	20.43	20.25
28	22.50	22.46	22.41	22.34	22.26	22.17	22.07	21.95	21.82	21.68	21.53	21.37	21.20	21.02
29	23.30	23.26	23.20	23.14	23.05	22.96	22.85	22.73	22.60	22.46	22.31	22.14	21.97	21.78
30	24.10	24.06	24.00	23.93	23.84	23.75	23.64	23.52	23.38	23.24	23.08	22.92	22.74	22.55
31	24.91	24.86	24.80	24.72	24.64	24.54	24.42	24.30	24.16	24.02	23.86	23.69	23.51	23.33
32	25.71	25.66	25.60	25.52	25.43	25.33	25.21	25.08	24.95	24.80	24.64	24.47	24.29	24.10
33	26.52	26.46	26.40	26.31	26.22	26.12	26.00	25.87	25.73	25.58	25.41	25.24	25.06	24.87
34	27.32	27.26	27.19	27.11	27.01	26.91	26.79	26.65	26.51	26.36	26.19	26.02	25.83	25.64
35	28.13	28.07	27.99	27.91	27.81	27.70	27.57	27.44	27.29	27.14	26.97	26.79	26.61	26.41

TABLE 10 (*cont.*)

S(‰)	T(°C)											
	14	15	16	17	18	19	20	21	22	23	24	25
0	−0.67	−0.81	−0.97	−1.14	−1.32	−1.51	−1.71	−1.92	−2.14	−2.38	−2.62	−2.87
1	0.10	−0.04	−0.20	−0.37	−0.55	−0.75	−0.95	−1.16	−1.39	−1.62	−1.86	−2.12
2	0.88	0.73	0.57	0.39	0.21	0.02	−0.19	−0.40	−0.63	−0.87	−1.11	−1.37
3	1.65	1.50	1.33	1.16	0.97	0.78	0.57	0.35	0.13	−0.11	−0.36	−0.61
4	2.42	2.26	2.10	1.92	1.74	1.54	1.33	1.11	0.88	0.64	0.39	0.14
5	3.19	3.03	2.86	2.69	2.50	2.30	2.09	1.87	1.64	1.40	1.15	0.89
6	3.96	3.80	3.63	3.45	3.26	3.06	2.85	2.62	2.39	2.15	1.90	1.64
7	4.73	4.57	4.39	4.21	4.02	3.82	3.60	3.38	3.15	2.90	2.65	2.39
8	5.49	5.33	5.16	4.97	4.78	4.58	4.36	4.13	3.90	3.65	3.40	3.14
9	6.26	6.10	5.92	5.74	5.54	5.33	5.12	4.89	4.65	4.41	4.15	3.88
10	7.03	6.86	6.69	6.50	6.30	6.09	5.87	5.64	5.41	5.16	4.90	4.63
11	7.80	7.63	7.45	7.26	7.06	6.85	6.63	6.40	6.16	5.91	5.65	5.38
12	8.56	8.39	8.21	8.02	7.82	7.61	7.38	7.15	6.91	6.66	6.40	6.13
13	9.33	9.16	8.98	8.78	8.58	8.36	8.14	7.91	7.66	7.41	7.15	6.88
14	10.10	9.92	9.74	9.54	9.34	9.12	8.90	8.66	8.42	8.16	7.90	7.63
15	10.86	10.69	10.50	10.30	10.10	9.88	9.65	9.41	9.17	8.91	8.65	8.37
16	11.63	11.45	11.26	11.06	10.85	10.63	10.41	10.17	9.92	9.66	9.40	9.12
17	12.39	12.21	12.02	11.82	11.61	11.39	11.16	10.92	10.67	10.41	10.14	9.87
18	13.16	12.98	12.79	12.58	12.37	12.15	11.92	11.67	11.42	11.16	10.89	10.62
19	13.93	13.74	13.55	13.34	13.13	12.90	12.67	12.43	12.18	11.91	11.64	11.36
20	14.69	14.51	14.31	14.10	13.89	13.66	13.43	13.18	12.93	12.66	12.39	12.11
21	15.46	15.27	15.07	14.86	14.65	14.42	14.18	13.93	13.68	13.41	13.14	12.86
22	16.22	16.03	15.83	15.62	15.40	15.17	14.94	14.69	14.43	14.17	13.89	13.61
23	16.99	16.80	16.60	16.38	16.16	15.93	15.69	15.44	15.18	14.92	14.64	14.36
24	17.76	17.56	17.36	17.14	16.92	16.69	16.45	16.20	15.94	15.67	15.39	15.10
25	18.52	18.33	18.12	17.91	17.68	17.45	17.20	16.95	16.69	16.42	16.14	15.85
26	19.29	19.09	18.88	18.67	18.44	18.20	17.96	17.70	17.44	17.17	16.89	16.60
27	20.06	19.86	19.65	19.43	19.20	18.96	18.71	18.46	18.20	17.92	17.64	17.35
28	20.82	20.62	20.41	20.19	19.96	19.72	19.47	19.21	18.95	18.67	18.39	18.10
29	21.59	21.39	21.17	20.95	20.72	20.48	20.23	19.97	19.70	19.43	19.14	18.85
30	22.36	22.15	21.94	21.71	21.48	21.24	20.99	20.73	20.46	20.18	19.89	19.60
31	23.13	22.92	22.70	22.48	22.24	22.00	21.74	21.48	21.21	20.93	20.65	20.35
32	23.90	23.69	23.47	23.24	23.00	22.76	22.50	22.24	21.97	21.69	21.40	21.10
33	24.67	24.45	24.23	24.00	23.76	23.52	23.26	23.00	22.72	22.44	22.15	21.86
34	25.43	25.22	25.00	24.77	24.53	24.28	24.02	23.75	23.48	23.20	22.91	22.61
35	26.21	25.99	25.77	25.53	25.29	25.04	24.78	24.51	24.24	23.95	23.66	23.36